TENSIONS IN TEACHING ABOUT TEACHING

Self Study of Teaching and Teacher Education Practices

Volume 5.

The titles published in the series are listed at the end of this volume.

TENSIONS IN TEACHING ABOUT TEACHING

Understanding Practice as a Teacher Educator

by

Amanda Berry
Monash University, Clayton, Australia

 Springer

A C.I.P. Catalogue record for this book is available from the Library of Congress.

ISBN-978-1-4020-5992-6 (HB)
ISBN-978-1-4020-5993-3 (e-book)

Published by Springer,
P.O. Box 17, 3300 AA Dordrecht, The Netherlands.

www.springer.com

Printed on acid-free paper

To Warren and Maxx

CONTENTS

PART ONE: CONTEXTS OF THE STUDY

CHAPTER ONE: Beginning to Research My Practice

CHAPTER TWO: Teacher Educators Studying Their Work

CHAPTER THREE: Developing a Research Approach

CHAPTER FOUR: Tensions as a Framework for Learning About Practice in Teacher Education

PART TWO: EXPLORING THE TENSIONS OF PRACTICE

CHAPTER FIVE: Telling and Growth

CHAPTER SIX: Confidence and Uncertainty

CHAPTER SEVEN: Action and Intent

CHAPTER EIGHT: Safety and challenge

CHAPTER NINE: Planning and Being Responsive

CHAPTER TEN: Valuing and Reconstructing Experience

CHAPTER ELEVEN: Revisiting and Summarising the Tensions

PART THREE: LEARNING FROM TEACHING ABOUT TEACHING

CHAPTER TWELVE: Becoming a Teacher Educator

ACKNOWLEDGEMENTS

Thank you to the Biology methods class (2001), at Monash University. Your willingness to participate in this study enabled me to expand the boundaries of my understanding of teacher education. Hopefully, as a consequence of your involvement, you will also have been able to see into, and pursue, new understandings of your own developing practice.

SERIES EDITOR'S FOREWORD

This series in Teacher Education: Self-study of Teacher Education Practices (S-STEP) has been created in order to offer clear and strong examples of self-study of teaching and teacher education practices. It explicitly values the work of teachers and teacher educators and through the research of their practice, offers insights into new ways of encouraging educational change. The series is designed to complement the International Handbook of Self-study of Teaching and Teacher Education practices (Loughran, Hamilton, LaBoskey, & Russell, 2004) and as such, helps to further define this important field of teaching and research.

Self-study of teaching and teacher education practices has become an important 'way in' to better understanding the complex world of teaching and learning about teaching. The questions, issues and concerns, of teacher educators in and of their own practice are dramatically different to those raised by observers of the field. Hence, self-study can be seen as an invitation to teacher educators to more meaningfully link research and practice in ways that matter for their pedagogy and, as a consequence, their students' learning about pedagogy.

Even a cursory glance at the literature illustrates that self-study has dramatically expanded since its inception in the late 1980s and early 1990s. Building on the foundations of fields such as reflective practice, practitioner inquiry and action research, self-study has continued to develop because, for many teacher educators, it has created opportunities for greater professional satisfaction in their teaching and research and, "Much of this work has provided a deep and critical look at practices and structures in teacher education" (Zeichner, 1999, p. 11).

Clarke and Erickson (2004), in reviewing shifts in views of teaching and learning over time, made clear the importance of relationships as a cohering theme for pedagogical development. They drew particular attention to the need for teachers to problematize practice and argued that in so doing, it helped to illustrate the importance of teaching and learning a site for inquiry. As they examined the links between what is learned and who does the learning, they worked towards a conceptualization of teachers as learners. Through their argument, they came to see self-study as a fifth commonplace, (as per Schwab's (1978) four commonplaces), describing it as the cornerstone of professional practice because, "without self-study teaching becomes repetitive not reflective – merely the duplication of models and strategies learned elsewhere and brought to bear unproblematically in one's own classroom" (p. 41).

Extending this notion of a fifth commonplace then, self-study can be seen as offering access to ways of knowing, or the professional knowledge of teaching and learning

about teaching as it is played out in the real world context of teacher education programs, teacher educators' teaching, and the learning of students of teaching.

In this volume, Berry offers a critical lens for viewing the work of self-study
through her remarkable, indepth, longitudinal study of her teacher education practices and do so in such a way as to more than illuminate the value and importance of
idea this fifth commonplace.

Through an exceptional array of data sources, Berry frames and reframes
(Schön, 1983) her practice in ways that highlight the analytic and methodological
rigour crucial to valuing the knowledge that emerges as a pedagogy of teacher education. She conceptualizes her practice through the notion of tensions and illustrates
not only how they arise, but also how they are 'played out', in ways that genuinely
shape her understanding of what she is doing, how and why in her teaching about
Biology teaching. There is little doubt that her articulation of these tensions is a powerful way of conceptualizing teaching and learning about teaching in ways that
might genuinely challenge, and therefore offer alternatives to, the "showing, telling,
guided practice" that Myers (2002) so rightly bemoans as the Achilles heel of some
teacher educator's practices.

In this book, Berry is concerned to make her approach to self-study open and
accessible to others in order to invite the critical review and debate so crucial to
scholarship and so central to advancing deeper understandings of the work of self-
study more generally. Her tensions of: Telling and Growth; Confidence and Uncertainty; Action and Intent; Safety and Challenge; Planning and Being Responsive;
and, Valuing and Reconstructing Experience, offer exciting ways of seeing into the
sophistication of her pedagogy of teacher education. However, moreso, these tensions are also a catalyst for others to reflect on, and pursue, ways of better articulating their own knowledge of practice and thereby further contributing to shared
understandings of a pedagogy of teacher education.

It is a great pleasure to be able to offer this superb example of self-study to readers of this series. It clearly breaks new ground in the field and makes clear for all to
see, why self-study of teaching and teacher education practices has such an allure
for teacher educators. I find the nature of the learning inherent in the study eluci-
dated by Berry through this book to be exceptionally compelling, I trust the case is
the same for you.

J. John Loughran
Series Editor

REFERENCES

Clarke, A., & Erickson, G. (2004). The nature of teaching and learning in self-study.
In J. J. Loughran, M. L. Hamilton, V. LaBoskey & T. Russell (Eds.), *International hand-
book of teaching and teacher education practices.* (Vol. 1, pp. 41–67). Dordrecht:
Kluwer Academic Press.

Myers, C. B. (2002). Can self-study challenge the belief that telling, showing and guided practice constitute adequate teacher education? In J. Loughran & T. Russell (Eds.), *Improving teacher education practices through self-study* (pp. 130–142). London: RoutledgeFalmer.

Schön, D. A. (1983). *The reflective practitioner: How professionals think in action*. New York: Basic Books.

Schwab, J. J. (1978). The practical: A language for the curriculum. In I. Westbury & J. Wilkof (Eds.), *Joseph J. Schwab: Science, curriculum and liberal education - selected essays* (pp. 287–321). Chicago: University of Chicago Press.

PREFACE

It is a common experience of teacher educators around the world that there is no formal preparation for their role as *teachers of teachers*. Many are successful former school teachers who have found themselves transformed almost overnight (Dinkelman, Margolis & Sikkenga, 2006) into their new situations. The nature of this transition is largely under-researched (Zeichner, 2005). As a consequence, many new teacher educators struggle to know what to teach their students about teaching and how to teach in ways that will effectively support their learning as new teachers.

The research reported in this book has arisen from my own struggles as a former high school Biology teacher and beginning Biology teacher educator learning to teach prospective teachers. The book is based on a substantial research project that aimed to explore, articulate and document the development of my knowledge of practice as a beginning Biology teacher educator. It outlines the development of my understanding of my pedagogy as a Biology teacher educator as I have made the transition from school teacher to academic.

In particular, I focus on the shared teaching and learning venture of teacher preparation through investigation of the experiences of prospective teachers in my Biology methods class learning about teaching, and myself, their teacher educator, learning to teach about Biology teaching. Through a self-study methodology (Hamilton, 1998), the development of my understanding of the importance of the relationship between my learning and that of the prospective teachers is explored, so that an articulation of my growing knowledge of teacher education practice begins to emerge.

CONTEXT OF THE STUDY

The research described in this book took place over one academic year (two semesters; March – October) within the Biology methods class in the Faculty of Education at Monash University, Australia. There are two teacher preparation pathways available at Monash: a one year Post Graduate Diploma in Education (Grad. Dip. Ed.); and a four year concurrent Double Degree (e.g., B.A./B.Ed., B.Sc./B.Ed.). Both programs prepare graduates for teaching in secondary schools. Students in the 4th year of the Double Degree program undertake their studies with Grad. Dip. Ed. students. Data sources for the research described in this book include myself, students in the Biology methods class within the teacher preparation program, and a colleague from the Faculty of Education.

The data sources were chosen in order to create genuine opportunities for me to see into my practice and prospective Biology teachers' learning from different perspectives to enhance critical reflection on my practice. Since the study of my practice also included tracing the sources of influence on my teacher and learner self, additional data were inevitably drawn upon to inform this study. These additional data included an autobiographical account of my self as a teacher and a learner, and an interview with a colleague about my teaching. The purpose of these data was to elicit and examine my beliefs about teaching, learning and the discipline of Biology and to trace the influence of these past beliefs and experiences into my current teacher education practice.

ABOUT THE ORGANISATION OF THIS BOOK

This book is organized into three parts. Part One: *Contexts of the study*, situates the study within my own teacher education setting and within the teacher education literature. The research approach and sources of data are outlined and the notion of tensions is introduced as the analytic frame that was developed through this research.

Part Two: *Exploring the tensions of practice*, examines each of the six tensions of practice that I came to identify through the research process, the development of my understanding of each tension and how each played out in, and influenced, my practice. This section concludes with a summary of the tensions and an exploration of the nature of the knowledge developed through conceptualizing practice through the frame of tensions.

Part Three: *Learning from teaching about teaching*, summarizes the learning about the tensions and brings them together into a frame for thinking about the development of knowledge of teaching about teaching for the teacher education community.

ABOUT THE PRESENTATION OF THIS BOOK

In this book, my voice as both researcher and researched is represented through the use of first person narrative. In this way, I explicitly recognize the role and contribution of the self (Hamilton, 2004) in self-study so that my teacher educator voice is overtly recognized and acknowledged. My approach offers an alternative to the ways in which traditional approaches to teacher education research have operated "at a distance from the practice of teacher education" (Zeichner, 1999, p. 12), and is consistent with the stance of feminist scholarship (Fine, 1992). Self-study research attempts to close the gap between the research and practice of teacher education through bringing forward the voice of the teacher educator, including her difficulties and vulnerabilities, as she reconceptualises not only her understanding of the practice of teacher education but also, what it means to engage in research as a teacher educator (Hamilton & Pinnegar, 1998a). I regard this as an important aspect of researching teacher education practices and a crucial element in reporting self-study.

LIST OF FIGURES AND TABLES

PART ONE

CONTEXTS OF THE STUDY

Chapter One

BEGINNING TO RESEARCH MY PRACTICE

When I began as a Biology teacher educator I brought with me many ideas about what I wanted prospective teachers to know about, in order to teach high school Biology, well. After a decade of teaching middle school Science and senior Biology in a variety of settings, I had developed a repertoire of successful pedagogical approaches, interesting activities and knowledge of 'what works' in the classroom. I also had been responsible for the school-based supervision of a number of prospective teachers during this time which led me to think that many of them seemed rather 'clueless' in terms of innovative approaches to teaching and learning. Hence, I believed that I had much to contribute to teacher education to improve the quality of high school Science teaching and learning.

However, although I brought much in the way of teaching experience to my new role, I brought little knowledge about the role of the teacher educator, other than my school-based supervisory experiences, and my own (distant!) memories of teacher preparation. This meant that in the teacher education classroom, I had few ideas about how to help new teachers learn about teaching other than sharing with them what I had done as a teacher. Initially, the approach I took to educating prospective Biology teachers involved just that – helping them to learn to reproduce my style. However, I soon realised that I could not simply expect prospective teachers to take on my approaches and my values in some unquestioned way. This realization created a sense of dissonance in me because, interestingly, the view of learning that I had developed over the years of teaching Science, (i.e., that students need opportunities to construct personally meaningful knowledge about Science with scaffolded teacher support), was not the view of learning that I was applying in this new teacher education setting. Not knowing how to act in the role of teacher educator led me to revert to an 'uninformed' model of learning, as in the "banking" model (Friere, 1986). It seemed that my professional knowledge of teaching had limited usefulness in terms of enacting a pedagogy of teacher education (Russell & Loughran, 2007).

This brief snapshot of my entry into teacher education parallels the experiences of numerous others. Yet, although the path from classroom teacher to teacher educator is one commonly traversed, surprisingly, it remains largely unmapped in terms of the development of professional knowledge of teaching about teaching. For me, these initial experiences as a teacher educator were confusing and challenging, particularly so, because I imagined that this new role was something that I would 'take on' and 'do', rather than one that I needed to construct and 'live'. I needed to better understand how to use what I knew to support the growth of others, although it took me some time to recognize that this was my task as a teacher educator.

This chapter then, sets out the motivation for the self-study described in this book – how my need to better understand my practice as a teacher educator grew and developed – and how this felt need led to the articulation of a series of research questions about my practice, and the subsequent emergence of a conceptual frame that serves as the foundation for this study.

Self-study (Hamilton, 1998) has been an important means of developing my professional self-understanding as a teacher educator, helping me to clarify what I bring to the role and how what I bring to teacher preparation may influence my actions and interactions with others in the learning to teach process. Through researching my practice I have come to recognize, articulate and (re)construct my pedagogy of teacher education. These experiences of researching understandings of practice have created a stepping off point for new growth and change.

DEVELOPING A RESEARCH APPROACH
THROUGH TRACING INFLUENCES ON PRACTICE

Bullough & Gitlin, (2001, p. 12) assert that "teacher education should begin with who the beginning teacher is – or rather, who you imagine yourself to be as a teacher – and then assist you to engage in the active exploration of the private or "implicit theories" (Clarke, 1988) you bring to teaching". As a beginning *teacher educator*, this seemed an equally important task for me, to explore who I "imagined [my]self to be" as a means of better understanding the relationship between my understandings of experience and my current beliefs and actions as a Biology teacher educator.

I began by tracing through my "education related life history" (Bullough & Gitlin, 2001). In so doing, I uncovered various assumptions and taken-for-granted beliefs (Brookfield, 1995) about teaching and learning that guided my work as an educator. For example, one assumption I held was that all students in my classes were motivated to achieve highly, just as I had been as a student; another assumption was that the process of learning is self-evident – one does not really need to learn how to learn. From these (and other) assumptions, together with issues and concerns that I also identified about my practice, I constructed a series of questions relevant to my current practice that could be pursued through self-study. Inevitably, as these initial assumptions were uncovered and investigated, new and more deeply embedded assumptions were brought to light, and further questions developed and explored. Figure 1.1, below, illustrates a selection of the questions that framed my initial approach to researching my practice.

Viewed together, the questions that I generated illustrated my ways of thinking about practice as a teacher educator that then guided the development of the major research foci of the study.

Additionally, I developed a 'Statement of Teaching Intentions' about my teacher education practices (see Figure 1.2). This document was expressed in

- What explicit and implicit messages about learners and learning do I convey through the manner in which I conduct Biology methods classes?
- Are these messages consistent with those I wish to develop in prospective Biology teachers?
- How much does it matter that the students in my classes like me?
- What assumptions do I make about how students approach learning in my classes?
- How can I create a methods course that acknowledges prospective teachers' histories as learners and is responsive to their needs, yet at the same time challenges their views and gives them the confidence and reason to try alternative approaches to teaching senior Biology (particularly when I have never experienced such a methods course, myself?)

Figure 1.1. Questions I Posed About My Teacher Education Practice

My goal as a Biology teacher educator is to assist the development of prospective Biology teachers who can develop ways of working with their students that will facilitate their meaningful learning (Ausubel, 1960) of Biology concepts and, that will stimulate their interest in and motivation to learn about Biology, both as a discipline and as a tool for understanding more about their own lives. I try to do this by constructing a learning environment that enables prospective teachers to experience the role of the teacher and of the learner and, to critically reflect upon the implications of these experiences for effective Biology teaching and learning for secondary school students.

My teaching philosophy has been influenced by a constructivist view of learning such that I recognise that knowledge is individually and actively constructed by learners on the basis of their experiences, values and attitudes. The process of knowledge construction is facilitated by social interaction, for example through shared experience and discussion. In science education, this view of knowledge construction is consistent with a social constructivist or Vygotskyian perspective in which, " . . . knowledge and understandings, including scientific understandings, are constructed when individuals engage socially in talk and activity about shared problems or tasks" (Driver, Asoko, Leach, Mortimer & Scott, 1994, p. 7). Social processes both enhance the meaning making process for individuals and at the same time provide a context for the creation of common meanings amongst a group of individual learners.

As a consequence of this philosophy, I believe that effective Biology teacher education involves:

- Accessing, identifying and building on prospective teachers' prior knowledge of learning, teaching and Biology concepts. One of the ways that I try to do this is by asking students to talk and write about their experiences as learners in Biology classrooms, and to encourage them to compare themselves with the students that they teach. Prospective teachers are often surprised to discover considerable differences between their own (high) interest and motivation and that of many of their students.
- Enabling prospective teachers to participate in, and be cognisant of, the process of exploring and constructing meaning.
- Creating situations that require prospective teachers to struggle with meaning making so that they might begin to recognise the time and effort required by their students to develop new knowledge.
- Exploring with prospective teachers the biological frames through which they see the world, and as a consequence, to begin to recognise the implications for teaching of the theory ladenness of observation.
- Helping student teachers to conceptualise Biology knowledge as a set of major interlinked, unifying principles that shape our understanding of our planet and of ourselves (Gess-Newsome & Lederman, 1995).
- Responding sensitively to the variety of ways in which learners experience and interpret the world and recognising the value of different interpretations in providing feedback to shape teaching to be better aligned with intended goals and purposes.

Figure 1.2. Statement of Teaching Intentions

terms of the broad goals for my teaching about Biology teaching and prospective teachers' learning about teaching, and a more detailed philosophy of teaching incorporating specific pedagogical aims, with examples from my practice. Careful scrutiny of this document provided additional insights into issues and concerns that further informed the research foci and approach. Figure 1.2 shows an excerpt from this document, outlining my broad goals, my teaching philosophy, specific aims for my teaching and an example from my practice to illustrate one of these aims.

My chosen approaches to investigating 'self' helped me to establish a framework for understanding my current practice so that I could monitor the effectiveness of my teaching against my teaching intentions and, at the same time, offered me a tangible entry point into researching practice. Taken together, the various questions, issues and concerns that I developed encapsulated the challenges and dilemmas of teaching about teaching central to a self-study approach to researching teaching about teaching. From a consideration of these, the conceptual frame for this research was developed and articulated.

ELABORATING A CONCEPTUAL FRAME

The major focus of this self-study is to articulate, document and analyse the challenges and complexities associated with my experiences as a Biology teacher educator learning to teach prospective teachers about teaching Biology. The conceptual frame outlined in Figure 1.3 represents the development of the research reported in this book, and is designed to illustrate how I have come to conceptualise my knowledge of practice of teacher education.

In constructing Figure 1.3 **bold type** is used in order to link the ideas of the schematic with the explanation that follows. Thus, the research developed through this self-study occurs in the **context of preservice teacher education**. The key participants are the **prospective teachers** in the Biology methods class and myself, their **teacher educator**. As a teacher educator I have particular concerns about how I teach and how my teaching about teaching might influence prospective teachers' learning about teaching. My concerns are based around ways of developing an **understanding of practice** that goes beyond the **technical aspects of practice** that preoccupy (at least initially) most prospective teachers' concerns about learning to teach Biology. These various concerns are framed as questions (outlined in Figure 1.3) for me, and for my students. **Concerns**, together with **needs** and **beliefs** are **factors that influence** both **learning about teaching** and **teaching about teaching**. For example, my pedagogical approach is strongly influenced by my belief that prospective teachers need to learn about teaching for themselves, rather than learning to reproduce another's style (either mine or that of another teacher/educator). This means that in my approach I provide considerable time and opportunity for students to try out, and to critique, new approaches to practice. However, the nature of the factors that influence my teaching about

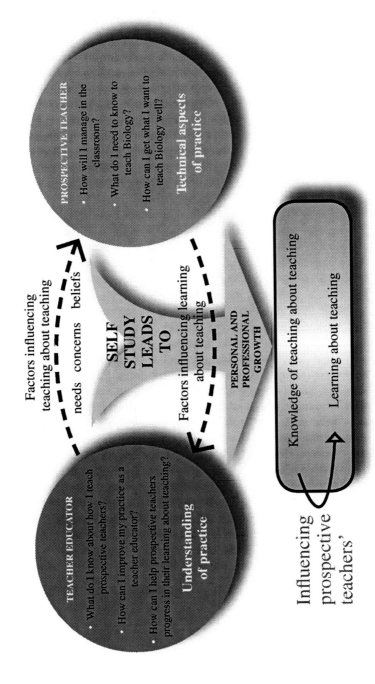

Figure 1.3. Schematic Representation of the Development of Knowledge of Practice of Teacher Education

teaching may be different from the nature of the factors influencing prospective teachers' expectations of learning to teach. For example, they may believe that learning to teach involves acquiring a set of 'tried and tested' classroom activities from me, their teacher educator. These differences in concerns, needs and beliefs between me and the prospective teachers that I teach, continually lead to shifts and adjustments in how we understand and enact teaching and learning. The interaction between these different factors is therefore represented as a cycle in Figure 1.3 and frames the way in this research is conceptualized and has been conducted.

Self-study is an approach to researching teacher education practice that is driven largely by the concerns of teaching and the development of knowledge about practice and the development of learning. In this research, self-study is the vehicle through which the nature of the relationship between my learning about teaching about teaching and prospective teachers' learning about teaching is explored and developed and which then **leads to personal and professional growth**.

The development of knowledge about practice through researching experience of practice brings together what is traditionally viewed as being somewhat separate and distinct in teacher education; i.e., a theory-practice divide (Wideen, Mayer-Smith & Moon, 1998; Korthagen, 2001). In fact, the notion of a theory-practice divide permeates much thinking about learning in teacher education in ways that are often unhelpful to the development and usefulness of each of these forms of knowledge. For example, knowledge about teaching presented in the form of theory has limited influence on learning about teaching because it is often disconnected from the learner's personal context hence it is not necessarily personally meaningful for the learner. This problem applies equally to teacher educators (learning about teaching prospective teachers) as it does to prospective teachers (who are learning to teach Biology to high school students). Thus, it is not difficult to see how the experiences of myself as a teacher educator and my students as beginning teachers offer access to ways in which this theory-practice divide impacts teaching and learning about teaching as we are both centrally situated in this study and simultaneously shaped by the resultant research.

Importantly, through a self-study methodology, learning about practice of all participants is continually facilitated and reinforced through a cyclical process of development. Therefore, as this process of investigating my practice has unfolded so my learning about teacher education practices has been informed. As a consequence, I have come to see the importance of a **knowledge of teaching about teaching** as a positive **influence on prospective teachers' learning about teaching**. And, that very process of learning connects participating individuals to a more elaborated understanding of a knowledge framework that in turn sets the foundations and expectations for personal and professional growth.

In the next chapter I explore in detail the literature that underpins this research in terms of the growth of knowledge of teaching about teaching that has developed through the self-study of teacher education practices.

Chapter Two

TEACHER EDUCATORS STUDYING THEIR WORK

For a long time in teacher education, we have heard the voices of educational researchers who do not burden themselves with the work of teacher education . . . but we have not often heard the voices of teacher educators themselves. Now we are hearing these voices in increasing numbers despite the unfavourable structural conditions of teacher educators' work. (Zeichner, 1999, p. 11)

Zeichner's (1999) observation that teacher educators have become increasingly involved in researching their own work heralds a new paradigm in teacher education research. For many years the perspectives and voices of teacher educators have been missing from educational research literature. This has meant that the concerns and needs of teacher educators about their work has received little serious attention since those involved in the study of teacher education were rarely involved in its day-to-day practices. Their research agendas were driven by different priorities and methodologies and produced knowledge about teaching and teacher education that was not necessarily helpful for the messy, context-specific problems faced by teacher educators, themselves. However, the failure of traditional paradigms in educational research to improve teacher education has paved the way for new forms of research to emerge, forms that more faithfully reflect the experiences and concerns of those who participate in it. This chapter provides a backdrop to the research presented in this book, situating it within the rise of the self-study movement and the development of interest in a pedagogy of teacher education. The chapter chronicles the motivations, approaches and learning of teacher educators engaged in researching their practices through exploration of the following questions: How do teacher educators develop their knowledge of teaching teachers? What informs the approaches they take? How do their chosen approaches affect prospective teachers' learning about teaching? What happens when teacher educators research their own teaching and, how does researching practice influence teacher educators' understandings of themselves, prospective teachers and the process of teacher education?

HOW DO TEACHER EDUCATORS DEVELOP THEIR KNOWLEDGE OF TEACHING TEACHERS?

Pathways of New Teacher Educators

Two pathways typify the entry of new teacher educators into the profession. One pathway leads from research, whereby researcher (as current research student, or newly conferred PhD) becomes teacher educator; the other pathway leads from classroom teaching, whereby successful teacher becomes teacher educator. However, describing these as 'pathways' into teacher education is a misnomer, since the term implies some sense of special preparation, or intentional career move, whereby intending teacher educators follow a structured path of learning about a scholarship of teacher preparation; a scenario that is, in fact, quite the opposite experience of most new teacher educators. The real situation is summarised well by Wilson (2006, p. 315) who says: "not . . . many scholars of this new generation have opportunities to learn to teach teachers in structured and scholarly apprenticeships; instead they are thrown into the practice of teacher education." Hence a major challenge for teacher educators lies in developing an understanding of their role in ways that are meaningful and helpful for the prospective teachers with whom they work (and that lead to effective student learning), particularly so when there is little in the way of ongoing professional support or mentoring (Zeichner, 2005; Lunenberg, 2002), or a well defined knowledge base of teaching about teaching (Korthagen, 2001).

The route via which they are jettisoned into their role impacts what new teacher educators bring to teacher preparation. On the one hand, those who have 'landed' as researchers, may bring much in the way of epistemic knowledge to impart to prospective teachers (although their research expertise rarely includes teacher education), yet little in the way of practical knowledge about teaching or an understanding of the current issues that face teachers and learners in schools (Zeichner, 2005). On the other hand, classroom teachers who move into teacher educator roles may bring considerable subject specialist expertise and a great deal of practical wisdom about dealing with the everyday realities of schooling, yet little in the way of theoretical understandings about teaching and learning. Because their knowledge has been developed within the practice context, classroom teachers often do not know how to offer what they know about teaching to prospective teachers in forms other than 'tips, tricks and good activities'. Unfortunately neither background is, in itself, particularly helpful for effectively supporting prospective teachers' learning about teaching, since teacher educators are required to play a "complex dual role" (Korthagen, Loughran & Lunenberg, 2005) that demands expertise both in teacher education research and in the kinds of skills and understandings that come from experience as a practitioner. This makes the role of teacher educator unlike that of their academic counterparts in other university faculties or professionals in other fields; teacher educators must both teach their subject area (i.e., teacher education) at the same time that they serve as role model practitioners for neophytes (ibid, 2005). Further, they must be able to articulate their pedagogy in ways that are comprehensible and useful for prospective teachers (Loughran & Berry, 2005).

Compounding these difficulties for new teacher educators are prevailing assumptions about teacher educators' work as a relatively straightforward task (i.e., that teacher preparation is a matter of 'simply' telling new teachers what they need to know), and unimportant within academia (compared with the more rewarded and valued tasks of research and grant writing). Hence, while new teacher educators may be well intentioned, they often do not recognise the complexities associated with their tasks, or that the knowledge they bring is insufficient for their new role. Two consequences of this situation are that, for a long time, the knowledge developed by teacher educators about their practice has remained static, tacit and weakly conceptualised (Berry & Scheele, 2007), and as an enterprise, teacher education has been more easily controlled by those outside the profession, since it is not well structured on the inside.

A growing number of teacher educators dissatisfied with these traditional 'plot lines' (Clandinin, 1995) of teacher education have been prompted to investigate its processes and their roles within it. They have resisted the forces compelling them to conform to traditional institutional norms and practices and instead have begun to construct new and different stories about teacher education (see for example, Guilfoyle, Hamilton, Pinnegar & Placier, 1995).

Developing Knowledge as a New Teacher Educator

Mary Lynn Hamilton's account of her initial experiences as a teacher educator reinforce the notion that, for a long time, there has been no specialized knowledge of teaching about teaching for teacher educators to draw upon.

> When I ask myself how I became a teacher educator, I am left puzzling about the first time I thought about doing that or left wondering if I ever really initiated a learning-to-be-a-teacher-educator process. I suppose though that I first began the process long before I became conscious of it. In the unconscious moments I worked hard to train teachers to integrate their curricula with multicultural perspectives or gender concerns. I spent long hours designing materials to be presented to teachers for use in their classrooms. But who taught me how to do that? Really no one taught me. I learned by watching those people around me, by reminding myself what happened in my own classrooms with high school students, by trying to remember the stages of development and how these might fit with what I needed to do. I also learnt by making errors, major errors in front of the classroom. No class at the university discussed the process of becoming a teacher educator. (Guilfoyle, Hamilton, Pinnegar & Placier, 1995, p. 40)

Two issues from Hamilton's account help to explain why a collective knowledge of teacher education has been slow to develop: one is that learning about teacher education is often experienced by the teacher educator as a private struggle, the other relates to the role of experience in developing knowledge as a teacher educator. Others have also reported these issues in their transition into teacher education

(Kremer-Hayon & Zuzovsky, 1995; Murray, 2005; Dinkelman, Margolis & Sikkenga, 2006). An individual, trial-and-error approach to learning about teacher preparation means that each teacher educator must 're-invent the wheel' in terms of learning to recognize and resolve problems encountered in the practice context. This is not to say that the development of knowledge through experience is not a worthy pursuit – it offers considerable potential for teacher educators' learning about practice – yet what is learnt is dependent on the skills and motivation of each individual teacher educator in how such experiences are analysed and understood. This is a point raised by Murray (2005, p. 78):

> Practical knowledge – developed in suitable settings, for worthwhile purposes, in appropriately reflective ways – can and should form an important part of what it means to be a teacher educator. However, if those conditions for the settings are not met, then that practical knowledge is in danger of becoming narrow, haphazard, technical and uninformed by a sense of the broader social and moral purposes of teacher education.

Developing one's knowledge of practice as a teacher educator in the absence of any structured support also leads to a tendency, at least initially, to reproduce practices experienced in one's own experiences of schooling or teacher education. Such re-enactment of past practices seems to occur whether or not the experience was regarded as helpful for learning (Kremer-Hayon & Zuzovsky, 1995; Ducharme, 1993). Interestingly, in institutions where support is offered, such as the setting in which Murray (1995) conducted her research, new teacher educators felt unsure about what support to ask for, since they did not yet know what they needed to know in order to progress in their roles (a situation that parallels that of many prospective teachers during their teacher preparation). Crowe and Whitlock (1999) offer an alternative perspective from their experiences as doctoral students and teacher educators. Both praised their faculty community as one that provided support and that valued experience and reflection on experience.

However, despite the difficult circumstances of their work, increasing numbers of teacher educators have become interested in better understanding and developing their knowledge of practice. Over the past decade, the study of teacher education by teacher educators themselves has moved from being a mostly private, ad hoc struggle to become a publicly acceptable academic pursuit. The American Education Research Association (AERA) Special Interest Group, Self-Study of Teacher Education Practices (S-STEP), created in 1992, (with a current membership of approximately 260), is testament to the acceptance by teacher educators (at least) of the relevance and value of examining the nature and development of their work with prospective teachers.

Self-Study of Teacher Education Practices

Self-study offers a means for teacher/educators to examine their beliefs, practices and their interrelationships (Hamilton, 1998). Self-study shares features with reflection and action research such that each involves identifying and clarifying 'problems of practice' and working towards deeper understandings of those problems and changed practice through planned and purposeful inquiry. Importantly, self-study builds on reflection as it takes both the individual and the 'problem' being studied beyond the level of the personal into the public domain to make what is learnt available to others. Through making self-study a public process, the knowledge and understanding that is developed can be "challenged, extended, transformed and translated by others" (Loughran, 2004, pp. 25–26). Also, while self-study may operate via the parameters of action research (for instance, using cycles of reflective inquiry) it is not restricted to these parameters. The manner in which self-studies develop is often more responsive to the given situation compared with a predefined problem-action cycle.

An important feature of the self-study process is that it "yields knowledge about practice" (Dinkelman, 2003, p. 9). The knowledge produced is intended both as a means of "reframing" (Schön, 1983) teacher educators' personal understandings of practice and stimulating the development of knowledge of practice amongst the community of teacher educators, more broadly. In so doing, self-study researchers aim to contribute to the knowledge base of teaching and generating new understandings of the world (Hamilton, 2004). A significant challenge for the self-study community lies in developing approaches to representing the knowledge produced by individual teacher educators that are consistent with the purposes of self-study and that can contribute to informing a pedagogy of teacher education.

The growth of interest and involvement of teacher educators in self-study has been supported by particular changes in the research climate over the past decade. Changes include increased attention to the concept of a profession and the knowledge base of professionals (how professionals 'know' and use what they know), growth in research methodologies that more faithfully represent the experiences of those who are portrayed in research (particularly women, and research employing feminist methodologies), and the development of forms of research that explore the particular pedagogical concerns, tensions and dilemmas that drive everyday practice (for example, action research and practitioner research). Changing conceptualisations about the nature of knowledge in teaching and learning have been important to the ways in which teacher educators have come to understand, describe and value their work.

Views of Knowledge and the Self-Study of Teacher Education Practices

Views of knowledge have traditionally been categorised as belonging to one of two different forms: knowledge that is propositional or theoretical, and knowledge that is experiential or practical. This dichotomous approach has led to the notion of a theory-practice divide. The separation between forms of knowledge has inevitably shaped the ways that knowledge has been organized, understood and valued in researching education (Munby, Russell & Martin, 2001). As a consequence, a pervasive and enduring tension exists within teacher education concerning the status accorded each of these forms of knowledge production and the usefulness of each form in the work of teaching.

Much of the knowledge produced about teacher education (and education more generally) has been reported in the form of theory and made available through a science-oriented research approach. Knowledge produced in this way is usually in the form of generalizations, or propositions, that are considered applicable to a wide range of context-independent situations (Korthagen & Kessels, 1996). Such forms of knowledge production have long been privileged within academia because they fit with academic ideals of technical 'elegance' and the pursuit of knowledge as 'timeless truths.' And, while knowledge produced in this way is intended for teachers (and teacher educators) to use, it has proved to have limited use for teachers because it does not recognize or respond to the difficulties associated with individuals' needs, concerns and practices. This is due to the fact that such knowledge is often stripped of the particulars of individual situations that are most relevant to the work of teaching. Teacher/educators want, and need, more practically oriented knowledge than what has traditionally been made available through empirically driven research. This is not to suggest that such knowledge is not useful, but to observe that it is not commonly made available in a form readily accessible to the practitioner.

In contrast to traditional forms of knowledge and knowledge production, practical knowledge is personal, context-bound, and gained through experience. It includes implicit knowing, that is, a kind of knowledge that is embedded within action that cannot be separated from that action (Eraut, 1994). Practical knowledge has not been accorded the same high status as 'traditional theoretical' knowledge within academia because the individual nature of what is learnt and how it is learnt does not conform to established paradigms of standpoint, validity and reliability. Despite this, the concept of practical knowledge has attracted increased attention by researchers looking to more faithfully capture the nature of experience in their work. A variety of constructs has been associated with the acquisition of such knowledge, including tacit understandings (Polanyi, 1966), reflection (Schön, 1983, 1987), authority of experience (Munby & Russell, 1992, 1994), nested knowing (Lyons, 1990) and reframing (Schön, 1983). Munby and Russell (1994) use the term "authority of experience" to capture the status of knowledge derived through personal experience, compared with other, traditional forms of authority such as the "authority of position" or the "authority of scholarly argument."

An important element of practical knowledge that is inevitably connected to the practice of self-study is self-knowledge. Acquiring practical knowledge involves the study of self and the notion of "putting the I in the centre of research" (McNiff, Lomax & Whitehead, 1996, p. 17). Central to this process is developing an increased awareness of how one's philosophy of teaching has been informed by the deeply embedded images, models, and conceptions from experiences as a learner (Brookfield, 1995) and the impact of these on teaching relationships with others.

Differentiation between knowledge types is apparent in the literature in many ways and to varying levels of specificity. For example, Fenstermacher (1994) differentiates between two types of practical knowledge: embodied knowledge or personal practical knowledge, exemplified through the work of Elbaz (1983) and Connelly and Clandinin (1985), and practical knowledge that is developed through reflection on practice, based on the work of Schön, and researchers who have built on Schön's work, including Munby and Russell (1992), Grimmett and Chelan (1990), and Erickson and Mackinnon (1991). Both types of practical knowledge, Fenstermacher argues, "seek a conception of knowledge arising out of action or experience that is itself grounded in this same action or experience" (p. 14). For self-study practitioners, conventional social science methods have been unhelpful for the development of understanding of practice; hence the search for new forms of representation that can capture the complex and personal nature of the knowledge acquired. Self-study has built on this development of alternative approaches to framing knowledge as the need for more appropriate and helpful conceptualizations for researching, understanding and describing teacher educators' work have been sought (see, for example, Carson, 1997; Korthagen, 2001; Fenstermacher, 1994). The work of Korthagen has, for many, been a useful way of revisiting these issues about knowledge and knowing, in his drawing upon the Aristotelian distinction between episteme and phronesis.

> *Episteme* can be characterised as abstract, objective, and propositional knowledge, the result of a generalization over many situations. Phronesis is perceptual knowledge, the practical wisdom based on the perception of a situation. It is the eye that one develops for a typical case, based on the perception of particulars. (Korthagen, 2001, pp. 30–31, italics in original)

Episteme and phronesis are useful constructs in understanding knowledge developed through teaching about teaching because they help to define the nature of the knowledge that is sought, developed and articulated both by teacher educators themselves and by the prospective teachers that they teach. However, simply categorizing knowledge differently does not necessarily reduce concerns about how knowledge influences practice for, as Korthagen further notes, "many teacher educators actually work from an episteme conception" (p. 29), even though they want that knowledge to be useable and useful to prospective teachers. This leads to teacher educators' ongoing dilemma of better aligning intentions and actions in practice, a dilemma that is often a catalyst for self-study. Korthagen sees promise in understanding the difference between episteme and phronesis, as he asserts that a better understanding of the interaction between both kinds of knowledge is important in the development of understanding of learning to teach others effectively. This kind of understanding is a crucial issue in self-study.

Munby, Russell, and Martin (2001) report "overwhelming evidence" to support the idea that knowledge of teaching is acquired through personal experience of teaching. Phronesis, then, offers an excellent means of conceptualising the knowledge developed through experience. It involves becoming aware of the salient features of one's experience, trying to see and refine perceptions, making one's own tacit knowledge explicit, and helping to capture the particularities of experience through the development of perceptual knowledge (Korthagen, 2001). It also involves selecting epistemic knowledge that links with particular contexts and situations to further make sense of experience, rather than imposing epistemic knowledge as the starting point. Korthagen's (2001) proposal for teacher educators "to help student teachers explore and refine their own perceptions . . . [by creating] the opportunity to reflect systematically on the details of their practical experiences" (p. 29) is also important in the process of knowledge development of teacher educators in their learning about teaching about teaching.

Teacher educators who engage in self-study may be viewed as responding to the development of knowledge as phronesis. Recognising the need to develop knowledge in this way does not automatically equip a person to do so, because holding knowledge in the form of phronesis requires both a collection of particular experiences and a grasp of generalities that arise from them. This means that inexperienced teacher educators, lacking a store of specific experiential knowledge to draw from and attempting to respond to traditional forms of research and knowledge, often find themselves in 'unchartered territory' as what they seek to know and their ways of coming to know are not always congruent. Phronesis links closely with Munby and Russell's (1994) notion of "authority of experience". An important consequence of viewing knowledge through the frame of phronesis is that perceptions of knowledge and its status change. The perceived privilege of traditional research knowledge is moderated, as it becomes only one part of the professional knowledge required for understanding practice.

Reconsidering different forms of knowledge and knowledge production in the light of episteme and phronesis frames traditional research as the production of epistemic knowledge and, practical inquiry as the investigation of phronesis. In many self-studies, teacher educators develop their phronesis as they learn how to make their knowledge available, practical and useful in their teaching about teaching. For some, investigating practice often begins by searching for knowledge about practice in the form of assumptions or taken-for-granted beliefs (Brookfield, 1995) that guide teaching actions. Practical inquiry aims to uncover such assumptions and to explore their effects in teacher educators' work. Often these assumptions elude investigation because they are so deeply embedded in an individual's approach. Brookfield (1995, p. 2) describes the process of assumption hunting as "one of the most challenging intellectual puzzles we face in our lives." He identifies the process of critical reflection as crucial to the assumption-hunting endeavour. Self-study involves locating one's assumptions about practice through the process of reflection, in order to facilitate the development of phronesis. Thus it appears that self-study involves developing knowledge as phronesis, understanding the conditions under which such knowledge develops, understanding the self, and working to improve the quality of the educational experience for those learning to teach.

Defining Knowledge Developed through Self-Study Matters

Teacher educators working to understand their own practice in their individual contexts may not necessarily be concerned with what kind of knowledge they are developing about practice, rather that they *are* developing a better understanding of what they do. However, examining the knowledge arising from self-study is important because if the efforts of individuals are confined solely to their own classrooms and contexts, the problems of teacher education will continue to be tackled individually and in isolation. In self-study, there is also a need to find ways to share what comes to be known in ways that are both accessible to others and that can serve as a useful foundation for the profession. This inevitably involves discussions of the nature of knowledge since self-study seeks to position teacher educators as knowledge producers, and therefore challenges traditional views of knowledge production as external, impersonal and empirically driven. When what teacher educators know from the study of their practice is able to be developed, articulated and communicated with meaning for others, then the influence of that might better inform teacher education, generally.

WHY ARE TEACHER EDUCATORS INTERESTED IN STUDYING THEIR PRACTICE? WHAT INFORMS THE APPROACHES THEY TAKE?

Teacher educators who engage in the self-study of their practices recognise teacher education as an enterprise that is fundamentally problematic by virtue of the complexity and ambiguity of its various demands. By researching their practice, teacher educators ask themselves about the problems of teacher education and question how their own actions contribute to these problems. Unpacking the complexity of teacher education through its sustained study has led to important insights about the unique nature of teaching *teaching,* compared with teaching other content, for example, social studies, psychology or working with special needs students. Such insights then, begin to illustrate that 'just being a teacher' in teacher education is insufficient to highlight the subtleties, skills and knowledge of teaching itself. Russell (1997, p. 44) identifies a "second level of thought about teaching" in teacher education that is "not always realized . . . one that focuses not on content but on *how* (author's italics) we teach" (ibid, p. 44). Loughran (2006) builds on this idea, explaining that how we teach involves more than modeling practices consistent with our messages to prospective teachers, it requires being able to articulate decisions about how we teach, as we teach, in ways that "gives students access to the pedagogical reasoning, uncertainties and dilemmas of practice that are inherent in understanding teaching as being problematic" (p. 6).

Developing an understanding of practice as making explicit that which is usually 'unseen' and, as a consequence unexamined, involves a shift in thinking about teacher preparation from a process of acquiring information and practising techniques to learning to recognize, confront and learn from problems encountered in practice. Viewing teacher education practice as a "learning problem" as opposed to a "technical training problem" (Cochran-Smith, 2004, p. 1) is an important indicator of this shift occurring and one that is closely connected with teacher educators'

motivations to study more closely the relationship between teaching and learning in
their work.

Motivations for Self-Study

Teacher educators engaging in self-study commonly share a broad motivation to
improve the experience of teacher education through improving their teaching
practice. Whitehead (1998) articulates this motivation to improve practice as a
series of questions: "How do I improve my practice?"; "How do I live my values
more fully in my practice?"; and, "How do I help my students improve the quality
of their learning?" Teacher educators who choose to study their practice also draw
on the idea of credibility as a motivating influence in their work. They ask themselves, "How can I be credible to those learning to teach if I do not practice what I
advocate for them?" Heaton and Lampert (1993) remind us that the credibility of
teacher educators is at risk if they do not use the practices that they envision are
possible for others.

Teacher educators' specific reasons for engaging in self-study vary and include:

Articulating a philosophy of practice and checking consistency
between practice and beliefs

Some teacher educators seek to better understand the various influences that guide
their thoughts and actions. From a more well-developed understanding of these
influences, more informed practice may result. For some teacher educators (particularly those new to teacher education/self-study), this may involve investigation of
their transition into their new role, so as to better understand and subsequently
shape, their developing identities as teacher educators (Dinkelman, Margolis &
Sikkenga, 2006; Ritter, 2006). For others, it may mean learning to articulate a philosophy of practice through investigating practice (see Nicol, 1997a). More experienced teacher educators may be prompted to explore the coherence between
philosophy and practice to uncover possible discrepancies between espoused
beliefs and the realities of practice (see Grimmett, 1997; Tidwell, 2002; Aubusson,
2006; Crowe & Berry, 2007). In a related study, Conle (1999) identified her need to
become more informed about aspects of her teaching practice that may have been
otherwise hidden from her view: "I undertook to study my teaching not because I
saw particular problems (I did see several), but in order to discover if there were
problems I did not see" (p. 803).

The desire to investigate practice can also be linked to a personal need to
ensure that one's teaching practice is congruent with expectations for prospective
teachers' developing practice. For example, although not explicitly identified as
self-study, Lampert identified the importance for her colleague, Heaton, of aligning her practice as a teacher educator more closely with her expectations for her
students' practice as teachers. Lampert observed: "the pedagogy of mathematics

she [Heaton] wanted to teach teachers differed from her own practice of teaching mathematics. She could not live with the dissonance" (Heaton & Lampert, 1993, p. 77). Through ongoing reflective examination of professional practice, thinking about teaching and teacher education is challenged and teacher educators' awareness of the influence of curricula and pedagogical decision-making is raised (Cole & Knowles, 1995).

Investigating a particular aspect of practice

Some self-studies are focused more specifically on the influence of a particular approach or task on prospective teachers' thinking about, or approach to, practice. For example, Holt-Reynolds and Johnson (2002) investigated artifacts of their practice (assignments for students) as a way of learning about prospective teachers' needs and concerns. These two teacher educators each developed assignments for their classes that were intended to provide opportunities for prospective teachers to work in different ways and to promote professional growth. Both teacher educators were puzzled to find that few students in their classes took up these opportunities in their assignment work. Through critical analysis of the assignment tasks they had set and their students' responses to these tasks, Holt-Reynolds and Johnson learned that prospective teachers' concerns about available time combined with habitual, ingrained ways of working outweighed their motivations to work differently. Other examples of self-studies investigating particular aspects of practice include Trumbull's (2000) analysis of the kinds of written feedback she provided on students' work and the congruency of her feedback with the messages about reflection that she was trying to promote, Mueller's (2001) study of the journal task she was using to promote reflection with prospective teachers and Brandenburg's (2004) study of the use of 'Round Table Reflection' as a means of enhancing critical reflection in her Mathematics methods classes.

Developing a model of critical reflection

Teacher educators seeking to make explicit to prospective teachers their pedagogical reasoning may use self-study as a means of monitoring their efforts. Heaton identified that "by making her teaching available for study to people who do not ordinarily engage in the careful analysis of actual practice . . . [she] makes available a situation in which the problems entailed in implementing those practices can be directly examined and understood from alternative points of view" (Heaton & Lampert, 1993, p. 46). Loughran's (1996) self-study of his modeling of reflection for his students and Hudson-Ross and Graham's (2000) investigation of the effects of modeling a constructivist approach in their teacher education practice are further examples of this type. Winter's (2006) self-study of her efforts to explicitly model and critique with prospective teacher-librarians her approaches to teaching, illustrates the considerable challenge associated with this task.

Generating more meaningful alternatives to institutional evaluation

Self-studies may be generated as alternative means of representing teacher educators' practice to their institution for purposes of promotion or tenure. Values
about teaching that are implicit in standard teaching evaluations may be at odds
with the kinds of values that teacher educators hold as most helpful for promoting
prospective teachers' learning about teaching. For example, teaching evaluation
questionnaires are often based on a 'teaching as delivery of information' model.
By choosing to evaluate practice through self-study, teacher educators may be in a
better position to more faithfully represent their intentions for practice to
others. The experiences of Fitzgerald, Farstad and Deemer (2002), belong to
this category.

An alternative way of categorising purposes for self-studies is according to the "levels of concern" that the study addresses (Hamilton & Pinnegar, 1998b). "Microlevels"
are local; they begin from the immediate context of the classroom and involve questions
such as, "How do I encourage participation of all students, rather than allowing a few to
dominate?" Self-studies that begin from "macrolevels" are initiated from more global
concerns such as, "Can I help promote social justice in schools through my work
with prospective teachers?" The self-studies compiled by Tidwell and Fitzgerald (2006)
illustrate well macrolevel issues of social justice, multiculturalism and equity.

Distinguishing and classifying different purposes for self-study is a difficult and
potentially misleading task. The nature of investigating practice is such that these
purposes cannot be easily categorized or 'held still in a spot.' The boundaries blur
because what is being studied offers insights into practice that then influence practice and inevitably, alter the focus of the study. Categorizing studies according to
purpose is also difficult because teacher educators rarely study one aspect of their
practice at a time; what is central at a particular time can move to the periphery as
other issues come to occupy the teacher educator's focus of attention. For example, a
teacher educator seeking to learn more about a particular teaching practice may be
led as a result of her enquiries to a more general investigation of practice, which may
lead to the uncovering of assumptions about teaching and the articulation of a philosophy and then back again to the original practice.

What this illustrates more broadly is that knowledge developed in teaching
about teaching usually emerges from teacher educators' efforts to solve "learning
problems" (Cochran-Smith, 2004). These problems may present themselves as
'surprises' encountered in the course of their work, or they may be the result of a
teacher educator's deliberate decision to investigate a particular aspect of
practice. Importantly, self-studies begin from inside the practice context, emerging from a real concern, issue or dilemma. In this way, a phronesis perspective of
knowledge development is demonstrated as teacher educators begin to apprehend,
describe and investigate their problems of practice. Through this process, better
understanding of the particular characteristics of individual contexts is developed, together with an appreciation of that which is unique to a pedagogy of
teacher education.

WHAT HAPPENS WHEN TEACHER EDUCATORS RESEARCH THEIR OWN TEACHING?

Pathways of Self-Study

While the term 'self-study' seems to suggest an exclusive focus on the teacher educator, the 'self' in self-study encompasses a more diverse variety of selves than the teacher educator alone. Inquiry into the nature of teacher preparation to better understand the experience of teaching prospective teachers can begin from a study of self where 'self' is the teacher educator, or through investigating an aspect of prospective teachers' experience where 'self' is the student/s. Alternatively, collaborative conversations with the 'selves' who are colleagues may serve as a starting point for the study of teaching about teaching.

Although the beginning points may be different, the 'selves' are intertwined in such a way that the study of one 'self' inevitably leads to study of an 'other'. For instance, teacher educators who begin by investigating prospective teachers' understanding of an aspect of their teacher preparation may be led to apprehend something about the nature of their own actions as a teacher and about the unintended effects of those actions. This, then, may set in motion an investigation of the teacher educator's own actions that were not part of the initial intention of the investigation. This is illustrated for example, in Dinkelman's (1999) inquiry into the development of critical reflection in preservice secondary teachers, a study that unexpectedly evolved into a powerful examination of Dinkelman's own teaching. By interviewing prospective teachers from his classes about their processes of reflection, Dinkelman came to learn that his own teaching approach was "unknowingly squelching . . . the most valued objectives of his teaching" (p. 2). He was drawn into a new kind of investigation of his teacher-self as a consequence of his willingness to listen to, and learn from, the prospective-teacher-selves who experienced his teaching.

In other studies, teacher educators intentionally begin from prospective teachers' experiences in order to access understandings of teaching practice that might otherwise be invisible to them. For example, Freese's analysis (2002) of a student's apparent resistance to reflect on his own teaching and Hoban's (1997) investigation of students' understanding of the relationship between his teaching and their learning are two self-studies in which the teacher educator deliberately sought to use prospective teachers' experiences as a mirror to look into personal teaching practice. Hoban described the reciprocal learning process that occurs when prospective teachers are asked to study their own learning, which then stimulates the teacher to study personal teaching practices.

Critical conversations with a colleague about her practice led Bass, a teacher educator, to scrutinize her own classroom interactions more closely (Bass, Anderson-Patton & Allender, 2002). Bass invited a colleague, Allender, into her classroom for a semester to give her feedback about her practice. Through the critical conversations they shared, Bass came to recognize 'points of vulnerability' in her approach to practice. Using this heightened awareness, Bass began to investigate how these vulnerable points were played out in her interactions with her students. The above shows that self-study is not a straightforward process, and this leads to a consideration of the ways in which learning from self-study is conceptualized.

SUMMARY: CONCEPTUALISING LEARNING FROM SELF-STUDY

Teacher educators have learnt a great deal that is worth sharing from the self-study of their practice. Their work makes a significant contribution to understanding and articulating a pedagogy of teacher education. However, for many teacher educators, capturing the learning associated with researching personal practice is a difficult task. Their difficulties lie not so much in recognizing their work as messy and complex (this is readily apparent to them), but in finding ways to represent the learning developed in such a way that honors the realities of practice in its messy complexity, and yet, is sufficiently meaningful and useful for a range of other readers. Addressing this issue has been a significant challenge for teacher educators and one that is taken up in this book through the notion of 'tensions of practice'. In the next two chapters, (chapters 3 & 4) these tensions of practice are introduced. Chapter 3 describes the research approach for the self-study reported in this book, introducing 'tensions' as an analytic frame and in chapter 4, these tensions are further elaborated and linked to the literature of self-study.

Chapter Three

DEVELOPING A RESEARCH APPROACH

SELF-STUDY AS A METHODOLOGICAL FRAME

This research focuses primarily on the development of my self-identity as a teacher educator. The self examined is both personal and professional, and includes my beliefs about teaching and learning and their possible sources, and my practice as a teacher educator, including my interactions with others in the context of my work. The methodological stance selected is that of self-study (Hamilton, 1998), as a means of examining beliefs, practices and their interrelationships. Pinnegar (1998) described the methodology of self-study thus:

> Self-study researchers seek to understand their practice settings. They observe their settings carefully, systematically collect data to represent and capture the observations they are making, study research from other methodologies for insights into their current practice, thoughtfully consider their own background and contribution to this setting, and reflect on any combination of these avenues in their attempts to understand . . . For these reasons . . . self-study is not a collection of particular methods but instead a methodology for studying professional practice settings. (Pinnegar, 1998, p. 33)

Hence, self-study as a methodology defines the focus of the study but not the way it is carried out (Loughran & Northfield, 1998). Instead, self-study draws on data sources that are appropriate to examining the issues, problems or dilemmas that are of concern to teacher educators. It is therefore common for such data to be drawn from multiple sources including discussions, journals and observations/recollections of practice (Loughran, Berry & Corrigan, 2001).

Gathering data from a variety of (primarily qualitative) sources is one of five principal characteristics of the methodology of self-study identified by LaBoskey (2004). The remaining four characteristics that typify self-study methodology, according to LaBoskey are, that the work is self-initiated and self-focused, improvement aimed, interactive (or collaborative) at one or more points during the process and, that validity is defined as a validation process based in trustworthiness.

Data Sources

The data for this study were developed in order to create genuine opportunities for me to see into my practice and prospective teachers' learning, from different perspectives. I needed to "stand in and outside myself" (Brookfield, 1995) to enhance critical reflection on my practice. Data sources included:

1. An autobiographical account of my experiences as a learner and teacher
2. Videotape of each of the two semesters of Biology methods classes that I taught during the one year period of the study
3. Two journals that I kept throughout the 2001 academic year (one public and one private journal)
4. Field notes that I took during Biology methods classes
5. Prospective teachers' responses to a 'Personal Learning Review' task (n = 28)
6. Interviews that I conducted twice during the year with a small cohort of prospective teachers from the class (n = 8)
7. Regular conversations with a colleague in the Faculty of Education
8. Regular e-mail correspondence between myself, and one of the prospective teachers in the Biology methods class, in which we explored our ideas about learning, teaching and Biology. (This data source emerged unplanned – a point that will be taken up in later in this chapter.)

Together, this comprehensive array of data sources contributed a rich picture of my practice through a range of different perspectives. The choice of each of these data sources and the particular methodological frame each offered for the study is now explained in more detail.

1. Autobiography Overview.   Life experiences, including the influence of social and cultural factors, shape teaching (Ball & Goodson, 1985). Pedagogical actions are often grounded in autobiographical experiences of learning, hence bringing these experiences to the surface can be an important step in coming to understand one's actions as a teacher/educator. Constructing life stories that portray the circumstances or choices that have led to a particular outcome is one important way in which these experiences may be made available for subsequent analysis (Bullough, 1996; Cole & Knowles, 1995).

Application within the study of my practice.   Prior to the commencement of formal data collection, I wrote an autobiographical account of my learning and teaching past. The purpose of this account was to identify and describe my beliefs and practices about teaching, learning and Science/Biology based on an examination of past experiences. I shared my account with a colleague in advance of a critically reflective conversation about my practice based both on my writing and my colleague's experiences of teaching with me. The purpose of our conversation was to uncover assumptions and pedagogical principles that guided my work with prospective teachers, to elicit examples and evidence of my beliefs in practice, and to search for contradictions and limitations in my writing as well as fuller explanations of my beliefs and practices (Brookfield, 1995).

My purpose for engaging in these autobiographical activities was: i) to produce an autobiographical narrative that established my pedagogical framework and hence would serve as the beginning point of my self-study; and, ii) to identify a set of assumptions about practice that I could use as a frame for analysis of my practice throughout the substantive data collection period. These insights and ideas provided me with a starting point for beginning to look more closely into my practice.

2. *Videotape Overview.* Mitchell and Weber note that dissonance between what one sees and how one feels, acts as a stimulus for reflection, thereby creating alternative possibilities for action (Mitchell & Weber, 1999). Videotape offers the opportunity to scrutinise practice through the recording and replay of classroom events (Mitchell & Weber, 1999; Harris & Pinnegar, 2000) that can lead to recognition of dissonance and to the re-creation of practice. While self-video offers a revealing view of oneself, it is not the whole of one's teaching experience, as Mitchell and Weber (1999, pp. 192–194) note:

> [Video is a] . . . nonetheless partial representation . . . based on what was visible or recordable via the camera's lens and microphone from a certain angle from a specific juncture in time and space. The video tape per se is not my view of myself. I watch the tape to experience and interpret this outside view of me, reconstructing or interrogating my self image in the process.

On the other hand, self-video can capture and make available for study, teaching actions and their effects. Actions and decisions that are taken 'in the moment' can be subsequently examined, and the full significance of particular actions can be pondered. So too, alternative approaches to practice may be imagined in order to achieve greater congruency between action and intent (Mitchell & Weber, 1999).

Application within the study of my practice. In this study, I video recorded seventeen of the twenty-one, two hour Biology methods classes that I taught during the 2001 academic year.[1] Classes were held in the same room each week, and the camera was set up in the same position for each session. One video camera was placed in the classroom, at an angle that could best capture the largest area of the space, particularly the parts of the room that I would be most likely to cover during my teaching. One area of the room was out of the camera view. This allowed a space for those students who did not wish to be filmed at all (though all students provided consent to be filmed) and for those who might temporarily wish to move out of the camera's gaze. Seeing myself through the 'eye' of the video camera confronted me with all of my practice, an experience that was challenging, as well as affirming. I used the videotape to help me obtain a 'slowed down look' at the events of each session and to become aware of aspects of my teaching that the video camera had captured, that I had otherwise overlooked. (For example, the videotape showed

---

[1] It was not possible to video record all sessions since some were conducted outside the university or before students had given permission for their involvement.

particular verbal/non-verbal responses from students that had gone unnoticed by me while I was teaching. Watching these afforded me greater insight into their experiences of the class.) Viewing the videotape also gave me an opportunity to look at myself 'from the outside' and to compare my feelings during the class, with how I saw myself on the screen. In this way, one important function of self-video for me was to see what I could not otherwise see, in my teaching.

3. Journals Overview.   Journal writing is commonly used in educational settings to assist the development of reflection. Journals can be used for a variety of purposes including data collection for documenting personal change, evaluating aspects of practice, to promote critical thinking, to release feelings and to develop observational skills (Ghaye & Lillyman, 1987). These purposes are equally applicable to teacher educators as to prospective teachers. Journal retrospection can be short and/or long term, looking back on immediate experience and/or the total experience. An important function of journal writing is that it provides the necessary distance and abstraction from the immediacy of teaching and, as such, serves as "a vehicle for reflection which then allows us to return to practice more thoughtfully, with, we hope, greater wisdom" (Adler, 1993, p. 163).

Application within the study of my practice.   I maintained two journals during the formal research period; one, a private journal, in which I recorded my thoughts, feelings and experiences of teaching Biology methods, and the other, an electronic journal, (I called this the 'Open' Journal), which was linked to the Biology methods Home Page within the Faculty of Education. My private journal included brief analyses of events as they arose within my practice, and my responses, in the form of brief comments, from viewing the videotaped recordings of Biology methods sessions.

The Open (electronic) Journal was publicly available to students studying Biology methods. This journal contained a record of my purposes for each session, how I saw these purposes unfold, as well as other observations that I made about my experiences of the class. An important purpose of the Open Journal was to provide prospective teachers with access to my thinking about the classes, including my aims, how I felt about whether or not these aims had been met, as well as other questions, concerns and observations arising from my experiences of the session. Students were informed that reading or contributing to the web page was not compulsory, but that they may find it useful to learn about my plans for their learning as well as being able to read about and give feedback (via e-mail or discussion) on our classes.

4. Field notes Overview.   Field Notes provide a written record of observations, interactions, conversations, situational details, and thoughts during a period of study. Different types of field notes are generated in the course of research, including mental notes, jotted (or scratch) notes and full field notes (Lofland & Lofland, 1995; Sanjek, 1990). In this study, jotted notes were the predominant type of field notes recorded. These consisted of brief "phrases, quotes, key words, and the like" (Lofland & Lofland, 1995, p. 90).

Application within my practice.   During Biology methods sessions I made brief hand written notes to myself about issues that I wished to raise or revisit with prospective teachers, comments that I overheard from prospective teachers about their experiences of sessions and summaries of notes made on the whiteboard or overhead projector trans-parencies. These field notes served both teaching and research purposes. I used them as prompts for discussion within the Biology methods class and informing the planning of subsequent classes (teaching), as well as acting as a means of recapturing the experiences of a session for my post class reflections (research).

5. Personal Learning Review Overview. Approaches to exploring 'self', such as life writing (Bullough & Gitlin, 2001) offer opportunities for prospective teachers to make explicit their thinking about schooling, teaching and learning. By making their ideas explicit, prospective teachers can begin to recognize the shaping forces and assumptions that influence their practice. When these artifacts of experience are shared with others, including teacher educators, common concerns and issues can be identified, discussed and individually, or collaboratively, investigated. Recognising the relationship between self and others is foundational to self-study. In this research, use of a Personal Learning Review (PLR) offered prospective teachers a way in to learning about themselves, at the same time that it encouraged the develop-ment of a deeper relationship (Bullough & Gitlin, 2001) with me, as their teacher educator.

Application within the study of my practice. The PLR task served both teaching and research purposes. The task consisted of a series of questions intended to gain a picture of prospective teachers' entering assumptions about Science, teaching and learning. The PLR was completed by all students as part of the formal requirements of the subject at the beginning of the academic year. The questions were intended as a stimulus for prospective teachers to begin to recognise similarities and differences between themselves as students/learners (e.g., what motivated them, what led to their success in Biology) and the students they taught. PLR responses were revisited via various activities and discussions throughout the year.

6. Interviews Overview. Interviews offer opportunities for detailed exploration of complex issues as participants give and receive information within a conversa-tional framework. Interviews provide a flexible approach to eliciting and exploring information from others since the interviewer can pursue a particular response with an individual, ask for elaboration or redefinition, or probe and pursue factors or feelings that arise during the exchange (Wiersma, 1986). A good interview enables access to the thoughts and feelings, as well as the knowledge associated with a particular experience of both interviewer and interviewee (Patton, 2002). In this study interviews were used as a means of gaining insight into others' experiences and understandings of my teaching, and their learning about teaching.

Application within the study of my practice. I conducted individual, audio-taped interviews with eight volunteers from the Biology methods class, twice during the year. The timing of each set of interviews was chosen so that a comparison could be made between prospective teachers' initial thoughts about teaching and learning and those they expressed as they prepared to leave the course. The intention of the interview on each occasion was to uncover prospective teachers' beliefs about them-selves as new Biology teachers, their views about learning, influences that shaped their teaching, and responses to their experiences of Biology methods classes.

7. Colleague interview and observation Overview.   Enlisting colleagues to share conversations about teaching offers insights into experience that are not possible when working alone (Brookfield, 1995). When colleagues, "listen to our stories and reflect back to us what they see and hear in them, we are often presented with a version of ourselves and our actions that comes as a surprise" (ibid, 1995, p. 141). Brookfield identifies five purposes for shared critical conversations with colleagues: (i) helps us to gain a clearer perspective on the parts of our practice that need closer critical scrutiny; (ii) increases our awareness of how much we take for granted in our teaching and how much of our practice is judgmental; (iii) can confirm privately felt instincts; (iv) suggests new possibilities for practice and new ways to analyse and respond to problems; and, (v) helps to break down a sense of isolation and as a result, to recognize commonalities of our individual experiences (op cit, 1995).

Application within the study of my practice.   I met regularly with a colleague to engage in (audiotaped) conversations about practice. Prior to the commencement of formal data collection we met for an autobiographical interview (see data source 1, earlier in this chapter) that explored various aspects of my educational beliefs, philosophy and practices. We continued to meet throughout the year to discuss different events from Biology methods sessions. In our discussions, my colleague acted as a critical friend, listening without judging, encouraging me to think further about my ideas, actions and feelings and prompting me to consider what these different events helped me learn about my practice and prospective teachers' learning about teaching Biology. At times, I felt considerably challenged by the questions my colleague asked me about my teaching decisions or actions. Although this was not always a comfortable experience, I found it valuable for my pedagogical growth.

On one occasion my colleague attended a Biology methods class as a participant/ observer. The purpose of the visit was two fold: as an approach to facilitating prospective teachers' understanding of their teaching through observation and discussion of mine using a third party (i.e., colleague) to facilitate this discussion; and, as a shared experience of my teaching, to further 'open up' conversations between my colleague and me about aspects of my teaching approach (particularly aspects that I was unable to recognize, at that point). After a brief introduction outlining the purpose for his visit to observe and critique my teaching with the group, my colleague participated in the activities of the class then led a debriefing discussion of my teaching at the conclusion of the session.

8. E-mail Overview.   Hoban (2004) identifies e-mail as a way of making personal insights public through communicating them with others and, because messages can be, "sent quickly . . . but downloaded when required, [e-mail] . . . is like having a 'slow motion conversation' because there is another layer of reflection when taking time to reply to someone's communication about their personal insights" (Hoban, 2004, p. 1047). Hence in teacher education, e-mail offers a form of information communication technology (ICT) that supports the development of thinking about, refining and reframing practice.

Application within the study of my practice.   In this study, regular e-mail conversations took place, over the academic year between myself, and Lisa (pseudonym), one of the prospective teachers in the Biology methods class. This unplanned data source emerged as a result of an experience early in the year when Lisa shared with me an extract from her personal, written journal that was a commentary on her experiences of the teacher education program. This led to a regular exchange of ideas between us, via e-mail and conversation, about our various experiences of teaching and learning. Through her questions and comments Lisa pushed me to consider why I taught the way I did, how I understood my interactions with others and, my beliefs about how learning can be authentic and meaningful for high school Biology students. My experiences with Lisa mirror those described by other teacher educators who have also developed e-mail relationships with prospective teachers in their classes for the purposes of facilitating understandings of teaching and learning (see Russell & Bullock, 1999).

Summary of data sources

To summarise, this study drew upon a broad range of qualitative data sources. Given the complex, diverse and richly detailed nature of professional practice, I needed a similarly rich and diverse set of artifacts that could help me understand more about my teaching from the perspectives of those that experienced it. Self-study research embraces the detailed nature of individual and collective experience and so employs methods that can best portray the inherent complexity of that work (Berry & Loughran, 2005).

Data Analysis

The emergence of tensions as a conceptual frame and analytic tool

My initial approach to data analysis was framed around the identification of that which I experienced as 'problematic' in my practice. Problematic situations were defined as those that caused me doubt, perplexity or surprise and that led me to question otherwise taken-for-granted aspects of my approach[2]. I identified and analysed critical incidents from within my practice (Measor, 1985) and investigated assumptions (Brookfield, 1995) that I held about teaching and learning. A further analytic approach involved exploring differences between my intentions for prospective teachers' learning and my actions in class. Preliminary analysis of the data developed alongside a review of the research published by other teacher educators also motivated to study their practice. As I read this literature I recognized similarities between my own experiences and what

---

[2] I use the word problematic in a Deweyan sense (1933) of problem as an intellectual difficulty.

these teacher educators reported from studies of their practice. I also saw a broader framework that connected the various elements of teacher educators' practice. It became apparent to me that teacher educators regularly experienced different *tensions* as they attempted to manage complex and conflicting pedagogical and personal demands within their work as teachers of prospective teachers. The notion of tensions seemed a useful way of describing teacher educators' experiences of their practice (including my own).

In addition, I recognised that the identification of particular tensions offered a useful conceptual frame that could be employed to organize and analyse studies of other teacher educators' investigations of their practice and, possibly as an analytic tool for the investigation of my own, derived from the themes and issues that had become apparent through the critical incidents, assumptions and behaviours that I had already identified.

In order to test the soundness of this frame as an analytic tool, I worked through samples of the various data sources, coding the data according to the different tensions. I picked out regular occurrences of the tensions at work within the data. This gave me confidence to commit to this concept as an analytic frame for researching my practice. The tensions therefore became both a conceptual tool for understanding teaching about teaching and an analytic tool for investigating my teaching about teaching. Given the purpose of my study as an investigation of my practice as a teacher educator in order to improve practice and prospective teachers' learning about practice, an interesting symmetry became apparent in the relationship between the teaching and research that comprised this study, which indeed, exemplifies the value of a self-study methodology.

Tensions identified

My understanding of the tensions developed as I worked with them. Through refinement, I established a final list of six. These include tensions experienced by teacher educators between: safety and challenge; action and intent; telling and growth; planning and being responsive; valuing and reconstructing experience; and, confidence and uncertainty. Each of these tensions will be elaborated in detail in the following chapter. My purpose here is to illustrate how the frame of tensions emerged through the research approach and, in this way, make explicit "the process of self-study" (Barnes, 1998), an important criterion for quality in self-study research (LaBoskey, 2004). Related to this, I now offer an example of one of the tensions, 'safety and challenge', together with two indicative data samples, in order to illustrate (briefly) the ideas of the tension and the manner of coding employed.

Illustrating a tension: safety and challenge
The tension between 'safety and challenge' emerges for teacher educators in engaging prospective teachers in forms of pedagogy that are intended to challenge ideas about teaching and learning, and pushing prospective teachers so far beyond their comfort zone that productive learning can not occur.

Data source: E-mail from Lisa (student)
Subject: Feedback about peer teaching
Date: May 21, 2001
From: Lisa
To: amanda.berry@Education.monash.edu.au
Have just read the open journal. I think it's funny to read some things from us
and from our own students. We try to wangle our way out of things that aren't
comfortable don't we? . . . I guess that came out in some of our comments – we
don't like drawing, less time for reflection. It gets uncomfortable when things
are a bit less structured than the norm.

Data Source: Interview with Ellie (student)

Ellie: . . . it's only been during this subject that I've actually put up my hand
and given my opinion . . . I've never felt safe to do that sort of stuff in
a classroom, like you'd be told you are wrong or that's a wrong opinion
to have. But you feel sort of safe in an environment where you can just
chuck things out there . . . It's sort of a safe place to make mistakes.
(Interview 1)

Each of the data sources (above) illustrates different aspects of the tension between
safety and challenge. The first example, an e-mail from Lisa, one of the prospective
teachers in the Biology methods class (see data source 8), reports her response to reading an Open Journal entry (see data source 3). Her e-mail follows a Biology methods
class in which prospective teachers were asked to draw, rather than write about, their
understanding of a concept, and to spend time reflecting on their experiences of learning and teaching. Lisa identifies a parallel between her own and her peers' response to
this situation, that challenged their expectations of teaching and learning, and that of
high school students when new ways of working are introduced. The tension then,
becomes apparent as one departs from the 'safety' of normal routines and how such
situations are managed and understood by the teacher/educator. The second example is
drawn from an interview with a prospective Biology teacher, Ellie, (see data source 6).
Ellie describes a new experience for her of feeling safe in a classroom. This in turn,
leads her to risk challenging herself to express her ideas publicly.

It is important to note that because of the complex nature of teaching and learning, I regularly found more than one tension embedded within a particular situation
or event. As a result, although the tensions may well appear neatly distinct from each
other in the presentation of this research, in reality, multiple tensions could be read
into individual instances.

SUMMARY

This chapter has outlined the research approach of the study described in this book.
The rationale for the methodological approach, the selection of data sources, the reasoning behind their inclusion and the process of analysis that led to the analytic

frame of tensions have each been described. Self-study, as a methodology, employs multiple methods, is self-focused and initiated, improvement aimed and exemplar based. The detailed manner in which the research approach has been described in this chapter is representative of a self-study approach that is concerned not only with generating knowledge of teaching, but also, aimed towards the improved under-standing of approaches to researching teaching about teaching. In the next chapter, I examine each of the tensions in detail, together with illustrative examples from teacher educators' accounts of their practice.

Chapter Four

TENSIONS AS A FRAMEWORK FOR LEARNING ABOUT PRACTICE IN TEACHER EDUCATION

Teacher educators' efforts to address problems of practice rarely result in tidy answers when such problems are viewed through the lens of self-study. Knowledge that is developed through teacher educators' investigations of their teaching about teaching reflects the "indeterminate swampy zone" of practice described by Schön (1987, p. 3). It is a complex and messy terrain, often difficult to describe. Grimmett (1997) found this when he attempted to capture the complexities of implementing a changed pedagogy in his classes: "I was to learn that, although there are solutions to some problems, every solution creates further problems in a classroom of diverse learning needs and expectations" (Grimmett, 1997, p. 131). Grimmett's words reflect the process of self-study itself, as a series of recursive spirals that lead to continuing investigations of practice.

For many teacher educators, the difficulties associated with researching personal practice lie not so much in recognizing the complexities inherent in their work (these they readily see) but in finding ways of representing that complexity to others. Because so little of the "swamp" has been mapped, it is hard to know how to proceed. It is important therefore, to bring together teacher educators' different accounts of their work to offer possibilities to others also wanting to learn to find their way around in that swamp of practice. Equally important is finding ways to represent these accounts in ways that preserve the complexity and ambiguity of the process of teacher educators' knowledge development yet, at the same time, are meaningful to the reader. Addressing this issue has been a significant challenge because teacher educators often learn from self-studies that they experience competing tensions, but they do not necessarily learn to articulate what those tensions are. That they are present within teacher educators' practice stands out clearly from individual self-study accounts in the literature, but these tensions are rarely organised or examined across studies to illuminate the patterns that exist. Invariably, these tensions do not present themselves neatly as well defined packages; rather, they interconnect. This chapter attempts to articulate and portray these different tensions in a way that makes them accessible to the reader.

EXPLICATING TENSIONS

The notion of tensions is intended to capture the feelings of internal turmoil that many teacher educators experience in their teaching about teaching as they find themselves pulled in different directions by competing concerns, and the difficulties for teacher educators in learning to recognize and manage these opposing forces. The idea of tensions is portrayed variously in research accounts as "deliberating about alternatives rather than making choices" (Nicol, 1997b, p. 96), "deciding which voices to listen to" (Brookfield, 1995, p. 45), and "conflicting stories" (Clandinin, 1995, p. 30). Loughran and Northfield (1998) identify tensions, together with disappointments and dilemmas, as "elements that dominate data gathering . . . [in self-study, that] occupy the study's centre of attention" (p. 14). Hence drawing on tensions seems an appropriate way of representing what happens in a self-study and how it shapes data gathering and the knowledge outcomes. The tensions that are described in this chapter include those that teacher educators have recognised in their own work as well as those that I have recognised from my reading of their work.

Many of these tensions have grown out of teacher educators' attempts to match their goals for prospective teachers' learning with the needs and concerns that prospective teachers express for their own learning. These, at times conflicting, purposes are part of the ever-present ambiguity of teachers' (and teacher educators') work; and are, as Lampert (1985, p. 194) observes, "more manageable than solveable". Tensions focus on the following areas, and although presented as a list, are not intended to represent a hierarchy. Each is elaborated in turn.

Telling and growth
- between informing and creating opportunities to reflect and self-direct
- between acknowledging prospective teachers' needs and concerns and challenging them to grow.

Confidence and uncertainty
- between making explicit the complexities and messiness of teaching and helping prospective teachers feel confident to progress
- between exposing vulnerability as a teacher educator and maintaining prospective teachers' confidence in the teacher educator as a leader.

Action and intent
- between working towards a particular ideal and jeopardising that ideal by the approach chosen to attain it.

Safety and challenge
- between a constructive learning experience and an uncomfortable learning experience.

Valuing and reconstructing experience
- between helping students recognise the 'authority of their experience' and helping them to see that there is more to teaching than simply acquiring experience.

Planning and being responsive
- between planning for learning and responding to learning opportunities as they arise in practice.

Telling and Growth

The first area of tension is embedded in teacher educators' learning how to balance their desires to tell prospective teachers what they know about teaching, with an understanding of the importance of providing opportunities for prospective teachers to learn about teaching, for themselves. The tension exists between informing and creating opportunities to reflect and self-direct, and between acknowledging prospective teachers' needs and concerns and challenging them to grow. Managing this tension is made all the more difficult by prospective teachers' strongly felt needs to be told 'what works' and by teacher educators' needs to be seen as helpful, thereby fulfilling traditional and subconscious perceptions of their role as teacher. Teacher educators often express this tension in comments such as, "How can I do my job and not come off as the only one in the class with all the answers?" (Pope, 1999, p. 178) or "How can I wean [prospective teachers] . . . from looking for recipes for good teaching?" (Adler, 1991, p. 77).

Both the teacher educator's role identity (Stets & Burke, 2000) and prospective teachers' expectations of teacher educators' behaviours as teachers can strongly reinforce the traditional 'telling' roles associated with teaching (Britzman, 1991). This creates a role dilemma for teacher educators seeking to enact new approaches to practice, as they are no longer sure about how and what to teach to be effective in their role. For example, Carson (1997, p. 78) experienced a role dilemma as he sought to free himself from "the trap of telling". He wanted to challenge this simplistic notion of teaching. Withdrawing the "authority of his experience" however, meant that he was left confused about how he should proceed as a teacher educator. He began to question what knowledge would be most helpful for prospective teachers if he could not tell them how to teach? "The students' frustrations were mirrored in the dilemmas that I felt in trying to negotiate the tension between informing students and creating opportunities for them to reflect" (p. 78). Through the study of his practice, Carson learnt to deal with his concerns about how to withdraw the authority of his considerable experience and instead use his experience to help prospective teachers to grow professionally through reflection on their own experiences.

Grimmett's (1997) experiences of learning to implement a pedagogy of inquiry reflect concerns similar to those of Carson.

> How do I step out of the role of presenting into the role of facilitating? How do I cast off the role of problem solving to engage in problem posing? How do I cease pouring energy into my performance as a teacher in order to channel it into meeting the needs of learners and monitoring the process of learning? (Grimmett, 1997, pp. 121–122)

Investigation of his practice led Grimmett to learn that being well-intentioned and knowledgeable about reflection as a teacher educator was not sufficient for addressing the complexities of learning to practice a reflective stance with his students. He needed to experience this first hand, in his own practice, and so consider anew through these experiences what it means to reflect.

Louie (2002) identified the emotional ties that can bind teacher educators to a particular belief about teaching and the difficulty of letting go of such ties, even when it is clear that they are not helpful for prospective teachers' learning: "I gradually became aware of my belief that lecturing was an essential element of good teaching . . . [and] I realised the discrepancy between my cognitive sense of good teaching strategies and my emotional tie to lecturing" (Louie, Stackman, Drevdahl & Purdy, 2002, p. 203). Her collaborative self-study led her to identify and better manage the tension between providing opportunities for students to develop independently and falling prey to her subconscious desires to fulfill the 'telling' role.

While the previous examples are drawn from the work of experienced teacher educators, new teacher educators also feel this tension. The collaborative action research undertaken by Dinkelman, Margolis, and Sikkenga (2001) explores the transition of Sikkenga and Margolis, two former high school teachers, into university-based teacher education. Each found himself experiencing the competing desires of wanting to tell prospective teachers in their classes about good teaching while, at the same time, each teacher educator acknowledged the importance for students of constructing this knowledge for themselves. Margolis came to recognise that his desire 'to tell' was a result of his "finally start[ing] to get good teaching, and you want your students to do the same thing" (p. 40). Telling is a powerfully seductive notion that can be extremely difficult to resist. It not only seems right; it is also easy to do. It is not surprising, then, to find that this first area of tension is well explored in the self-study literature, particularly given the prevalence of the transmission model in teaching and the mounting research evidence to suggest its limited impact on learning.

Tensions associated with teacher educators' attempts to build an environment that encourages prospective teachers to actively direct their own learning processes are further intensified in contexts where formal assessment systems are imposed. What would motivate prospective teachers to seek their own solutions to teaching problems when their formal assessment is at stake? Prospective teachers anxious to learn what they must do in order to be academically successful may be reluctant to risk sacrificing grades to respond to their real needs and, as a consequence, reinforce the 'teacher as informer' role. The background experiences of 'typical' teacher education candidates tend to support this scenario (Britzman, 1991). Given these circumstances, teacher educators find that their attempts to recognise and respond to the particular concerns of prospective teachers can be extremely challenging. Tidwell's (2002) self-study articulated these difficulties, as she investigated the question of how she attempted to incorporate valuing individual students' ways of knowing with "institutional standards and institutional norms" (p. 31). As a consequence of her self-study, Tidwell came to the unexpected finding that her *own* beliefs tended to limit the ways in which *she* valued differences between individuals.

A further aspect of this tension for teacher educators (particularly in the USA) is the influence of individual accountability, tenure and formal evaluations of teaching on perceptions of teaching. Fitzgerald, Farstad and Deemer (2002) described the challenge of enacting an interactive, learner-centred model of teaching, while being held formally accountable for their teaching based on "an instrument developed for linear teaching ('teaching as telling')" (p. 208). Fitzgerald recalled the conflicting feelings that the end of year student reviews evoked in her.

> While members of my promotion and tenure committee were supportive, rarely did they fail to point out the poor ratings by students on some items of the student evaluation survey. Uncertain if my interpretation of the ratings would be convincing, I dreaded seeing the numbers come in, and became anxious about their presence. At the same time, I resisted changing my practice in ways that might lead to higher scores on items which presume teacher dominance in the classroom. (Fitzgerald, Farstad & Deemer, 2002, p. 214)

Fitzgerald and her colleagues learnt to manage this tension by reframing the criteria for promotion and tenure to include data from the self-studies of their practice. They were successful in illustrating for their institutions the value of self-study for examining the interactive forms of practice they valued.

While in some situations, negative evaluations of teacher education approaches that seek to challenge the 'telling' model can hinder teacher educators' employment opportunities, in others, prospective teachers can be reluctant to provide critical evaluation of their teacher educators' practice because they do not want to disadvantage their own formal assessment. This situation became apparent for Hoban (1997) in his study of prospective teachers' reflection on their learning about teaching from their experiences in his classes. Both Hoban and his students reported the difficulties they felt in critiquing his teaching and their learning, honestly, knowing that their efforts would contribute to their formal assessment. Hoban learnt from his self-study that his students needed to trust that he valued constructive criticism before they were prepared to engage in it. Ungraded teacher education courses (or ungraded subjects offered within teacher education programs) offer some relief from the pressure to conform to role expectations for teacher educators and teacher education students. Nevertheless, it would be naïve to think that grading is the only obstacle to honesty and the pursuit of genuine, personal understanding in learning about teaching.

A further strand of the tension between 'telling and growth' lies in acknowledging prospective teachers' needs and concerns and challenging them to move beyond these. Nicol (1997b), for example, investigated her teaching about Mathematics teaching and recognised that her desire to teach teaching in such a way that encouraged and enabled prospective teachers to reflect on the purposes and consequences of their actions, conflicted with the expectations of many of the prospective teachers in her classes. (Their concern was to be told how to teach mathematics and what mathematics to teach.) Through her self-study, Nicol's perceptions of the balance, "between accomplishing . . . [her] own teaching goals and experiencing teaching from prospective teachers' eyes" (p. 112) were sharpened. She learnt to "reframe" (Schön, 1987)

this tension in terms of the differences between introducing her own agenda and
responding to prospective teachers' individual needs.

Noddings (2001, p. 103) summarises the tension embedded in 'telling and
growth' well, through the following: "I do not think the tension between shaping
students toward some preestablished ideal and encouraging them to grow in directions they themselves choose can be resolved. It is a tension that has to be lived".

Confidence and Uncertainty

As teacher educators begin to explore new ways of working with prospective teachers, many begin to experience feelings of self-doubt and uncertainty about what their
role entails. This leads to a second area of tension, between making explicit the complexities and messiness of teaching and helping prospective teachers feel confident
to develop as new teachers. Similarly, there is a tension between exposing one's vulnerability as a teacher educator and maintaining prospective teachers' confidence in
the teacher educator as a competent leader.

Teacher educators report feelings of uncertainty as they begin to enact new
approaches to practice. These feelings can be conveyed to student teachers who may
interpret them as a shortcoming on the part of the teacher educator. Deciding what
aspects of practice to make explicit, how to make them explicit, and when, so that they
are useful and meaningful for prospective teachers, lies at the heart of this tension. It is a
risky business for the teacher educator and requires the establishment of a trusting relationship with the class, as Hoban (1997) Loughran (1996) and Winter (2006) each learnt.

Clandinin (1995) invokes the analogy of competing authorities and different 'stories' to describe the professional risk associated with new approaches to practice.
Teacher educators who "give up a familiar and privileged story for the uncertainty of
a new one" (Clandinin, 1995, p. 30) can find the consequences extremely challenging,
both personally and professionally. White (2002) found this as she attempted to
implement a new teaching 'story' with her elementary Mathematics methods classes.

> Finding myself in the middle of a class peopled by students and content, I was uncer-
> tain what specific actions to take that might be constructivist in nature, or when to
> take them. Knowing what not to do did little to nothing to inform me about what to
> do. My teaching was analogous to trying to walk on quicksand. I had no lodestone
> from which I could launch my teaching to begin to establish a foundation from which
> to operate. Most of the students in the elementary maths methods class became frus-
> trated with me saying I was unclear and did not provide adequate leadership or direc-
> tion. Frustration for them translated into anxiety for me. (White, 2002, p. 308)

Another teacher educator, Schulte (2001) was able to build new understandings
of her practice through learning to manage the ambiguous notion of being confident
about uncertainty, and coming to see its value.

> I was insecure and doubtful, but this study also led to a certain confidence. Forc-
> ing myself to "risk" my relationships with students so that I might challenge

them to better understand multiple perspectives has provided me with a base of experiences to draw upon in the future. My students have said that many of the strategies and activities I used were successful, at least in the short term, in helping them to challenge their assumptions about teaching and themselves. I was often scared and anxious about my behaviors that were intended to disrupt students' thinking; however, I feel a little bit more prepared for the next time I will have similar interactions. Practice and my students' positive feedback have given me courage. (Schulte, 2001, p. 109)

Teacher educators who choose to share authority with their students expose their limitations, which can lead to a shared vulnerability that students may be very unwilling to accept. As Lampert (1985, p. 193) identifies, "Thinking of one's job as figuring out how to live with a web of related problems that cannot be solved seems like an admission of weakness". Teacher educators need to exhibit confidence so that prospective teachers can trust, and then risk, different ways of working. Examples of beginning teacher educators' self-studies highlight the difficulties of trying to establish oneself as a teacher educator (particularly when one is unsure of how to do so) and trying to provide a credible and convincing model for prospective teachers at the same time (see Carson, 1997; Peterman, 1997; Brandenburg, 2007).

Action and Intent

A third area of tension arises from the approaches chosen by teacher educators to bring about change in their practice and prospective teachers' learning about practice; between working towards a particular ideal and jeopardising this ideal by the approach chosen to attain it.

Much of what is learnt by teacher educators from the self-studies of their practice connects to the realisation that the goals they set out to achieve often are inadvertently undermined by their own choice of actions to achieve them. Senese (2002), in his study of his efforts to hand over responsibility for their learning to the students in his high school British Literature classes, sums up this dualistic notion as 'the attraction of opposites'. Senese's self-study resulted in an understanding that in order to free his students to be the independent learners that he hoped they would become, he had to set boundaries for their learning. This approach seemed contradictory to his instinctive predispositions. While Senese's work was predominantly concerned with high school students, the same ideas about the 'attraction of opposites' apply equally well to the way in which the learning of adults is conceptualized and organized, as the following extract suggests.

I had long believed that my primary job as a teacher was to make my students independent of my instruction and of me. I had strongly subscribed to providing students with multiple opportunities to learn, with choices, and with creative outlets. But as often as I turned the curriculum over to the students, I had still maintained control of it, doling out pieces as I saw fit, gauging how much was good for them, and allowing them to move forward only in measured steps. (Senese, 2002, p. 44)

In a similar example, Grimmett (1997) wanted to create a situation that engaged prospective teachers in his classes in free-flowing discussion, but soon learnt that he could not do this simply by letting discussion flow freely. He recognised that, "there could not be an *equitable* distribution of student voice when I, as teacher, was not creating the structures and opportunities for *equal* student access to the classroom discourse" (p. 129).

These examples from Senese and Grimmett focus on the contradictory effects of each individual's conscious actions to influence student learning. In a different example, Macgillivray (1997) highlights the ways in which practice can be sabotaged by one's unconscious beliefs, as she unwittingly undermined her own efforts to create more equitable discourses in her classes, "reinforcing much of what . . . [she] had attempted to disrupt" (p. 469). By researching her practice, Macgillivray identified her subconscious assumptions about power structures that served to distort her best intentions for prospective teachers' learning and that caused her to work "within and against myself" (p. 470). Although Macgillivray does not name her work as self-study, (she calls it "turning my philosophical stances inwards to see the contradictions in myself" p. 470), her research offers important insights for self-study practitioners about the powerful, shaping effects of unconscious assumptions on one's practice. Similarly, Tidwell's (2002) research, described earlier, is a further example of uncovering unconscious biases in practice that work against beliefs. An outcome of these self-studies is that each teacher educator identified the tacit rules that guided her interactions with others. These rules related to perceptions of effectiveness and success as a teacher educator, in order to play the role of teacher in certain ways. Interactions with prospective teachers that did not fit comfortably with these self-conceptions and prospective teachers' willingness to explore the reasons for "interactional misfires" (Macgillivray, 1997, p. 479) created opportunities for each teacher educator to acquire new self-knowledge.

Brookfield (1995) suggests that teacher educators need to question the assumptions and practices that seem to make their teaching lives easier but actually work against their long-term interests. However, before teacher educators can begin to question assumptions, they must recognize that they exist. Questioning a familiar and comfortable practice becomes much more fruitful after realizing that it is counterproductive. When particular patterns of behaviour become habitual, they come to be thought of as 'natural' and 'self-evident,' even though they may be working against the intended goals for others' learning (Wilkes, 1998). With many demands on their attention, teacher educators may not readily see the ways in which they themselves may be contributing to the 'opposites' effect, even though they may be readily apparent to their students or colleagues. Wilkes (1998) identifies an example from her own practice, recalling prospective teachers who struggle and seek help (a situation familiar to many teacher educators). Her example also illustrates the first area of tension, between telling and growth, thus highlighting the recursive nature of categorisation.

Often when a student comes to me for help, and they are truly struggling, my intuition tells me to help them either by giving them the answer or telling them where

to find it. It is painful for me to listen to them struggle and not give them the information they need. I often have to resist mightily what I want to do, what my gut
tells me, and fix the momentary crisis. But I have learned that if I become the
source of answers, then I often enable students to stop searching for themselves.
So I now employ what, for me, is a counterintuitive practice. I just ask them questions instead, such as, 'Why do you think it is important to know this?' . . . Later
they often come back and thank me for not telling them the answer. But at the
time, they often leave angry with me for withholding information from them.
(Wilkes, 1998, p. 199)

Teacher educators' realisation of the need to work in ways that are counterintuitive, and the problems of so doing, point to a growing area of self-study literature.
The difficulties that may be encountered as a result of working in different ways can
test the relationship between teacher educators and prospective teachers. This issue
is explored in the following tension.

Safety and Challenge

A fourth area of tension (named by Korthagen, 2001) comes from the process of
engaging prospective teachers in forms of pedagogy that may be perceived as
challenging, even confrontational, and acting with sensitivity and care as a
teacher educator. New approaches to teaching about teaching encourage opening
up one's practice to the scrutiny of others through honest discussions about the
impact of teaching on the development of others' learning. Inquiry conducted into
practice in this way confronts the usual 'rules of politeness' that guide discourse
amongst prospective teachers and teacher educators. Working with prospective
teachers in ways that genuinely open up practice for honest critique requires a
sensitive appreciation of others' feelings, such as the "caring" described by Noddings (2001).

Berry and Loughran's (2002) work with prospective teachers provides insight
into this tension as together they attempted to set up opportunities for students in
their classes to experience and to articulate the uncertainties of practice as they
encountered them through microteaching situations. Berry and Loughran sought
ways to help prospective teachers 'see into' their practice. Sometimes they did this
by confronting students with problems or possibilities in the moment of teaching –
an approach they acknowledged as risky, given the vulnerability of the prospective
teachers in this context.

Making decisions about which approach to take, with whom, and what aspect of the
teaching to highlight is risky and it cuts both ways. Not just [prospective teachers']
self-esteem was at stake, so too was our credibility as teacher educators. Students
need to know that we genuinely care about them. It is imperative that we do not
belittle or humiliate them, but, at the same time, we want them to feel uncomfortable

enough about their practice to begin to examine the implications of their teaching decisions and actions. (Berry & Loughran, 2002, p. 21)

Within the field of self-study this tension tends to be hinted at more than explicitly examined. Shulte's (2001) work suggests why: teacher educators often find it difficult to ask difficult questions of their students because their sense of identity is bound closely to their ability to develop good relationships with students. Teacher educators may feel that challenging prospective teachers' views may compromise this aspect of their role identity, hence they resist. (Interestingly, although issues associated with the use of confrontational pedagogies are frequently discussed by teacher educators at self-study conferences and are clearly felt within their work, these discussions, or the episodes that give rise to them, are rarely transformed into print.) Schulte (2001) explains her experiences of this role dilemma:

> Engaging students in this kind of confrontational pedagogy was a challenge for me, because my self-identity is often closely tied to my ability to relate to others. Jordan (1991) explains this in saying that a woman's deepest sense of being is continuously formed in connection with others. . . . Because I am continuously weighing the consequences of my actions on my relationships, assisting others in transformation is even more stressful for me. (p. 7)

Schulte's efforts to induce prospective teachers' self-examination and critical questioning led to struggles in her teaching that she had not anticipated. Just as Berry, Loughran and Schulte highlight the effects of confrontational pedagogies on prospective teachers' learning, so too Guilfoyle et al. draw on Piaget's notion of disequilibrium to describe what takes place when learning is disturbed.

> Our students perhaps are seldom faced with 'real' learning so they do not know how to deal with the disequilibrium and take it out on us. Most were good students and did not have to struggle. Why should they have to struggle with ideas now[?] (p. 194)

Valuing and Reconstructing Experience

Helping prospective teachers to understand that learning about teaching requires more than acquiring experiences of teaching, sets up a fifth area of tension. This tension is felt in helping prospective teachers recognise the "authority of experience" (Munby & Russell, 1994) that they bring, and have, during teacher education, and helping them to see that experience itself is insufficient a teacher about teaching.

The notion, "authority of experience" expresses the significance of the knowledge that individuals develop as a consequence of their personal experiences. Such "authority of experience" is often undervalued or ignored by teacher educators compared with other forms of authority, such as research texts or the teacher educator's own "authority of position" (Munby & Russell, 1994). Developing ways to both acknowledge and extend prospective teachers' "authority of experience" is the focus of Loughran and Russell's (1997) collaborative examination of their different programs of teacher education. Loughran and Russell identified the importance of "meeting students on their own terms" through valuing the ideas and experiences that prospective teachers bring to teacher education and "challenging them to interpret their own meaning in ways that they have not had to before and to translate insights into future teaching" (p. 164). The pedagogical challenge in this for teacher educators, and the source of tension, comes from developing approaches that do more than simply (re)confirm students' existing beliefs, so that they may be prepared to willingly suspend their beliefs, and to consider alternative approaches to pedagogy.

There are few examples of self-studies that clearly illustrate this tension in action. Brandenburg's (2007) work illustrates one teacher educator's efforts to move prospective teachers beyond the knowledge developed through their experiences of teaching into new kinds of knowledge and new ways of understanding practice. In her research, Brandenburg and prospective teachers in her Mathematics methods classes used each other as 'pedagogical sounding boards', sharing personal experiences of teaching so as to encourage each other to identify and make sense of the knowledge gained through experience.

Planning and Being Responsive

A sixth area of tension emerges from learning experiences that are pre-planned and unplanned "teachable moments" (van Manen, 1990; Hoban & Ferry, 2001) that arise within the practice context. Powerful learning for both teacher educators and prospective teachers can occur from unplanned moments but for this learning to occur, participants must be open to understanding the learning situation from the point of view of the learner, rather than imposing predetermined frames for learning.

An example of this tension is illustrated through Nicol's (1997b) self-study. Nicol identified the difficulties she experienced as a teacher educator having particular goals and intentions for prospective teachers' learning yet, at the same time, wanting to be responsive to the kinds of issues and concerns raised in her classes. Through the study of her practice, Nicol became aware of the delicate balance between listening *for* her teacher educator agenda and listening *to* what prospective teachers were really saying about their learning to teach experiences. She came to recognise that "a focus on only listening *for* makes it difficult to listen to students' experiences . . . [and] a focus on only listening *to* may make it difficult to interpret students' experiences" (p. 112). Learning new ways of responding to prospective teachers in her classes was an important outcome of Nicol's study. For her, this came from learning "to see practice through others' eyes" (Loughran, 2002, p. 33) so that she was not just responding according to her own preplanned agenda.

Heaton and Lampert (1993, p. 56) identify what they see as the requirements of teacher education that operates in a responsive manner: "Teaching *teaching* for understanding requires the teacher educator and the learner to interact in the context of actual teaching problems and to try to understand these problems in terms of the circumstances in which they arise". Discovering when to let go of a prior agenda in order to respond to prospective teachers' needs as they arise through experiences in teacher education is something that Pope (1999) also began to better understand through her self-study. She recognized that part of the process for her involved letting go of her own defensiveness, to shift from thinking about herself, and towards viewing situations from the perspective of her students. In situations such as this, rather than prescribing and controlling the learning experiences, creating conditions so that learning can occur becomes much more important (Northfield & Loughran, 1996, p. 126).

UNDERSTANDING PRACTICE AS 'TENSIONS TO BE MANAGED' AND THE POSITION OF TEACHER EDUCATORS

Although represented here as separate, the tensions that influence learning about practice in self-study do not exist in isolation of each other. They interact in practice in ways that produce a 'whole that is greater than the sum of its parts'. Interconnections bring to light new understandings and hence, new kinds of knowledge of practice. The ambiguities and complexities inherent in teacher educators' work often become more apparent through their investigations of practice, as they come to recognise themselves as "living contradictions" (Whitehead, 1993) and as they learn to become more comfortable with "build[ing] a working identity that is constructively ambiguous" (Lampert, 1985, p. 178). In fact, what is frequently learnt from teacher educators' self-studies is the importance of acknowledging, living within, and even embracing ambiguity in one's work. Instead of constructing tensions in a way that evokes despair and frustration, as problems to be solved or to be eliminated from one's work, teacher educators begin to "reframe" (Schön, 1983) tensions as elements that are necessary, even enjoyable, for the growth and learning that they bring.

One way of viewing this reframing process is through the concepts discussed in Chapter 2, of episteme and phronesis. Teacher educators engaged in self-study often struggle with the frustration that they may know what changes they wish to make to practice and possess the formal knowledge to support their reasoning (episteme) but do not have the personal, experiential knowledge (phronesis) to carry out their role in the manner they wish. Learning how and when to enact their knowledge or indeed what kind of knowledge to bring to bear in a particular situation remains a central focus of teacher educators' ongoing self-study efforts. Through such research one important purpose of sharing knowledge of practice becomes clearer: In recognising aspects of practice that have been identified by others, teacher educators may begin to be more sensitive to what is happening within their own teaching about teaching and become aware of possible alternative ways of responding. Consequently, the value of learning through framing and reframing involves being more informed about actions, intentions and interpretations of practice. (Later in this book, I argue further the importance of the concepts of episteme and phronesis as a frame for understanding the knowledge developed in teaching about teaching.)

SUMMARY

In this chapter, I have articulated the frame of 'tensions' as a means of conceptualising the problematic nature of teacher educators' work and as an approach to articulating knowledge of teaching about teaching. Analysing practice through the frame of tensions is one way of making explicit teacher educators' professional knowledge. In Part 2, I use the frame of 'tensions' as a conceptual tool for analyzing my own practice as a Biology teacher educator and articulating the professional knowledge that comprises my pedagogy of teacher education.

PART TWO

EXPLORING THE TENSIONS OF PRACTICE

Chapter Five

TELLING AND GROWTH

tell (v) relate or narrate in speech or writing; make known; instruct

grow (v) germinate, sprout, spring up, come naturally into existence

> The image of teaching as telling permeates every move we make as teachers, far more deeply than we would ever care to admit to ourselves or others. (Russell, 1999, p. 222)

INTRODUCTION

This chapter explores my experiences of the first of the named tensions; that is, the tension between *telling and growth*. I investigate the ways in which I tried to manage competing feelings of wanting to tell prospective Biology teachers what I thought they needed to know about teaching while, at the same time, supporting their growth as they learnt about teaching for themselves. I learnt that the tension between telling and growth exists as two interrelated strands: one is that of teacher educators providing information to students about teaching and creating opportunities for students to reflect and self-direct; and the other, is in teacher educators acknowledging students' needs and concerns and challenging them to grow beyond these immediate preoccupations.

The tension between telling and growth hinges on an acceptance that telling is most commonly an attempt to transfer propositional knowledge (a set of assertions which might apply generally to many different situations) from the teacher to the student and, that although such transfer may occur, it does not carry sufficient understanding to the receiver of the information to be personally meaningful or useful. Notions of teaching as telling (and learning as listening and remembering) are deeply embedded cultural 'myths' (Britzman, 1991). As McDiarmid (1990) observes, ". . . beginning teacher education students believe that teaching subject matter involves telling or showing, the view of teaching prevalent not only in schools but in the broader culture" (p. 13). One consequence of these myths is prospective teachers' expectations that they will be told how to teach in their teacher preparation, so that they, in turn, may teach their own students by 'telling' them. Such a view of teaching is often perpetuated by teacher educators and teacher education programs through program structures or selected teaching approaches. Teacher educators who choose to challenge this view of teaching as telling, as I did, face a complex and considerable task. Their task includes rethinking how one behaves as a teacher educator if the role of authoritative 'teller' is removed. In the remainder of this chapter, I examine the different threads of this tension; how I became aware of them, prospective teachers' responses, and how I grappled with issues that arose as a consequence of this tension at work within my practice.

INFORMING AND CREATING OPPORTUNITIES TO REFLECT AND SELF-DIRECT

My Views of Learning and the 'Telling' Model

My views of teaching Biology are informed by conceptual change and constructivist theories of learning (Gunstone, 2000; Posner, Strike, Hewson & Gertzog, 1982). My views of *teaching about teaching* Biology teaching have many features in common with constructivism and conceptual change learning in Science. That is, in order for meaningful learning to occur (as opposed to the passive accumulation of facts) teachers need to appreciate the individual and diverse range of conceptualisations of particular phenomena held by their students and to structure classroom experiences in such a way as to make new ideas intelligible (i.e., they need to make sense and be understood), plausible (i.e., they seem to be reasonable things to do) and fruitful (i.e., they can be used successfully by students, producing desired results) (Posner et al., 1982). Therefore the pedagogical approach I employed both in the design of the Biology methods curriculum and in working with prospective Biology teachers was intended to promote an understanding of learning as intellectually active construction of knowledge and to assist the development these new teachers as reflective practitioners, who are committed to examining the models of teaching that guide their work to inform ongoing professional learning and improvement. Prior to the commencement of the research, I identified some basic principles that guide my work as a teacher educator. These included the following:

- Learning to teach does not mean learning to teach *like me*. Being an effective teacher educator means that I need to develop ways of working that are responsive to and encourage the strengths, interests and concerns of individual preservice teachers rather than their learning to reproduce my approach.
- Knowledge is individually and actively constructed by learners on the basis of their experiences, values and attitudes. The process of knowledge construction is facilitated by social interaction, for example through shared experience and discussion.

It is clear from these principles that my intended emphasis was on nurturing the development of students as individuals. Therefore, the construction of the Biology methods subject and the way I worked with Biology methods students was purpose-fully organised to reflect these ideas.

Planned Teacher Educator Behaviours to Encourage Growth

I employed particular pedagogical approaches in my classes in order to create an environment that supported the growth of prospective teachers as developers of their own understandings of teaching. My intentions were to encourage students to facili-tate their own growth, discuss their own and each other's ideas and experiences and to reduce their expectation that I will tell them what they need to know to be suc-cessful. The pedagogical approaches I employed included increasing "wait time" (Rowe, 1974 a & b), withholding/delaying judgment and encouraging increased 'student to student' discussion, rather than acting as the conduit for all classroom talk. I now briefly explain each of these approaches and illustrate their use with some examples from my practice.

(1) Wait time: Wait time refers to the period of time a teacher remains silent after posing a question (wait time I) and the time following an initial student response (wait time II) (Rowe, 1974). Typically, teacher wait times (I and II) are very short (~ 1 sec). Lengthening wait time (3 to 5 seconds) improves significantly the number and length of student responses, students' speculative thinking and contributions from a variety of students. Because students generally find the idea of wait time intelligible and plausible (i.e., it makes sense to them and they can see that it produces the desired results of increasing responses) I purposefully modeled wait time in the first few weeks of the course and explicitly drew prospective teachers' attention to what I was doing and why. For example, I noted the follow-ing in my journal after watching videotape from the second week of methods class:

> Good wait time. I waited and then Warren came in with a thoughtful response. I said to the class, "What did you notice about how long I waited?" Then I gave an explanation of why wait time matters. Wish I'd asked them more about how that felt for them. (Personal Journal, Week 1)

However, even when I was aware of the importance of modeling wait time, I did not consistently model this behaviour. For instance, later in the same class, I again noted that:

> I cut that discussion off without any wait time. Just after talking about wait time with them too! (Personal Journal, Week 1)

Interestingly, in talking to Nick (prospective teacher) during our first interview, this same classroom "episode" (White, 1988) of wait time and its impact on stu-dents' responses was raised. I asked Nick what he thought were the main messages about teaching and learning that he understood from Biology methods classes so far. He looked a little unsure about my question, so I said to him, *"It's okay I'll give you some wait time."* He immediately picked up on the cue of wait time and replied:

Nick: Well, that's one thing. The importance of wait time. . . . I think it was about the second lesson. You mentioned it after you did it and it actually worked. Like you asked a question and I mean I'm the kind of person who gets the answer in my head straight away or doesn't get the answer ever and I'm sort of thinking, what's she doing? And someone actually answered and you explained why that was and it worked out well with students.
(Nick Interview 1: 193–200)

It is clear that even though I prompted Nick to think about wait time, he did recognize its purpose and he saw that one purpose of waiting (to increase the number of students responding) was achieved. Nick also recognized that although he didn't think that wait time was an effective strategy for improving his thinking, he did see that it impacted on others in this situation; an opportunity for learning that may have been less convincing had I simply told him about it.

A different occasion in which I became aware that another of the prospective teachers was beginning to grapple with ideas about wait time, followed a peer teaching episode[1] (week 15). I asked Sue (prospective teacher) to lead the peer teaching debrief. In her approach, Sue showed a keen interest in encouraging a range of students to contribute to the discussion. At one point, she asked for responses from those who hadn't yet made a contribution.

Sue: Can we get one more question from someone that hasn't said anything? (Pauses – she looks a bit worried, no one has responded)
Mandi: It's okay. Just give them some wait time. (Long pause)
Sue: How long is wait time? (Pause)
Pat: I've got a question . . . (I give non verbal signal to Sue to indicate, 'See, waiting works!')
(Videotape: Week 15)

In this "brief but vivid moment" (Mason, 2002), the tension between telling and growth is illustrated well as Sue feels what it is like to experience wait time in a way that is quite different to me 'telling' her (or the rest of the class) about it. She has had a chance to build her own knowledge by being in the situation of learning about wait time *by waiting*. This episode also illustrates the nature of teaching about teaching as recognizing opportunities for new growth. In this situation, I recognize and act upon a potential learning moment for Sue (and possibly other class members too) as it arises; embedding learning about wait time within a real experience. I encourage Sue to step out and take a risk with me as we explored this notion of wait time together.

---

[1] Peer Teaching involved students, in pairs, designing a mini-lesson, teaching it to their peers then debriefing the teaching and learning experience, afterwards.

(2) Withholding/delaying judgment: This refers to the teacher's actions in establishing and exploring a range of student views, opinions and explanations, rather than the teacher 'closing' on a correct response. The teacher's primary purpose in withholding or delaying judgment is to have as many students as possible clarify their views, reasons for holding them, and any differences between their views and those of other students (Baird & Northfield, 1992). In so doing, opportunities for learners to become aware of, reconsider and possibly restructure their meaning for ideas become available.

In Biology methods classes, I employed a range of tactics in order to withhold and delay judgment. For example, I used encouraging but non-committal responses such as "uh-huh", or "thanks" following student responses; I repeated what students said using a questioning tone as a means of checking an idea with the class; I asked for responses from non-contributors to a discussion and, I tried to pick out from the variety of responses those points which seemed to have been resolved and those requiring further clarification. While my intention was to encourage prospective teachers to rely less on me as an arbiter of their progress and encourage more interaction between themselves and their ideas to develop their thinking, this was not necessarily how my actions were interpreted in practice. For example, Lisa described my approach as "aloof". She found that she was trying harder to elicit a positive response from me, to please me, because I wasn't offering praise or 'judging' contributions, even though she recognized the intent behind my approach. Lisa's experience prompted her to begin to consider more closely how she might teach to encourage self-reliant behaviour in her own students. Her e-mail provides evidence of her thinking.

Date: Tues, 13 March 2001
From: Lisa
To: amanda.berry@education.monash.edu.au

I can sense that Mandi is very wary of the teacher pleasers in our class. I see that she tries to avoid responding to them by being a bit aloof. I will try this. She also doesn't give many judgments about what people say. Perhaps this has the opposite effect from that intended – i.e. students are trying to please her, but she doesn't seem pleased – will have to try harder. R [another lecturer] is less aloof and probably a bit more encouraging and judging about discussion and responses – she actually say 'yes, great' quite a lot. Mandi doesn't do this nearly as much. Interesting that I feel no need to please R. I think I have some kind of answer – try to be encouraging to everyone and let everyone feel they have 'pleased' me. I can be selective in how I do this and offer praise when students do things for themselves, so that they learn to derive their own pleasure from their own learning.

Even though I believed the goal of my non-judgmental approach was worthwhile, I struggled to know, at times, how I might help students move forward in their thinking from collecting ideas to making sense of them. Baird & Northfield (1992) highlighted the difficulties associated with this approach, as ". . . the teaching skills associated with this technique are complex and subtle; the teacher is making decisions continuously about whether and how to react to student comments as well as whether and how to close or redirect the discussion" (p. 232). Also, students' expectations of themselves as learners and their view of teaching impact considerably on their view of the worthwhileness of such an approach. In the Biology methods class those prospective teachers who did not expect to value their own ideas often saw little of worth for their own teaching from this aspect of my approach.

(3)Increasing student to student interaction: This refers to instances of increased direct student to student interaction, without the teacher acting as a conduit for, and/or commentator about, each response. In traditional patterns of classroom interactions the teacher poses a question, one student responds, the teacher comments on that student's response, another student responds, the teacher comments, and so on, as though each student contribution is only valid when sanctioned by the teacher, for example by agreeing, praising, rewording, redirecting, etc. When the teacher's controlling/judging role is withdrawn, there are increased opportunities for students to put forward and clarify their own thinking and to take greater responsibility for their own learning. This is not to say that the teacher does nothing; simply standing back to watch the students talk. There is a more complex and subtle role for the teacher to take in knowing when, how and if to intervene.

In the second Biology methods class (Week 1) I observed a tendency of mine that, during discussions, I commented on each student's response after it was offered. I told the class I was concerned that my behaviour perpetuated an expectation that I would control all discussion and that their responses needed to be somehow approved by me for discussion to proceed. At the time, the students seemed rather puzzled by my observation, as though teachers commenting on student contributions was an unproblematic 'given' of teaching. Privately, I chided myself that it was perhaps too early to make explicit (and problematic) to the class this aspect of my teaching since most of the prospective teachers had not yet even had a chance to run a discussion, let alone critique the teacher's role within it.

I did not explicitly pursue this matter with the class, although I resolved to continue to address this aspect within my practice. I was surprised then, when one week after I had raised the issue of teachers' involvement in discussion, one of the prospective teachers, Josh, told me that he had been thinking about what I had said. He admitted that up until that point he had just assumed that this was how teachers behaved during discussions. He had never questioned why teachers acted this way, or what messages might be conveyed through their actions. Hearing Josh's thinking surprised and delighted me. He had begun to critique the role of the teacher in discussion.

Reflecting on my conversation with Josh also helped me recognize an important aspect of the tension associated with helping prospective teachers learn about teaching. Creating conditions for growth does not necessarily mean that growth occurs immediately or obviously. It involves waiting and trusting that those conditions will be helpful for students' learning – over time. Just because these prospective teachers hadn't picked up on my intentions for their learning when I expected them to, did not mean that my intentions were inappropriate or unrealized. Josh had taken something away from the class and had kept turning it over in his mind. I had assumed that they weren't ready to deal with these ideas because I couldn't see them doing it on-the-spot and in front of me.

FURTHER ASPECTS OF MY APPROACH INTENDED TO STIMULATE INDIVIDUAL GROWTH

In addition to the specific teaching approaches already described, my overall approach to teaching Biology methods was based on creating an environment intended to support prospective teachers' personal knowledge growth, rather than me telling them what (I thought) they needed to know. In order to create such an environment, I used experiential learning situations that could act as a starting point for their learning. For example, in my approach to helping students learn about planning lessons effectively (an aspect of learning to teach that is commonly presented to prospective teachers as expert knowledge to be acquired and practiced), I began from their early experiences of planning and implementing lessons during their Science camp[2] (tasks for which they had had little formal prior knowledge input). I asked students what they had learnt about planning lessons from their experiences of teaching year 7s (12–13 year olds) on camp. Then, drawing from their ideas we considered together the purpose of a lesson plan and the different ways in which these purposes might be addressed. In my web-based Open Journal, I wrote about the power of this session as an example of experiential learning.

> I found it [session] really interesting . . . because people had already had experiences of planning lessons . . . they had a lot to contribute about the difficulties and important things to remember in lesson planning. It reminds me (again) that better learning comes from drawing on people's experiences, rather than simply telling people what to do. ("This is a lesson plan, this is how you make one".) (Open Journal Week 3)

I hoped that from collaboratively constructing knowledge about planning through reflection on practical experiences that the development of their understanding about the purposes for planning lessons might be enhanced. In this way, lesson planning might then become a more meaningful idea to them compared with (what is often perceived by them as) yet another (meaningless) university requirement. The experience of teaching year 7s on camp had hopefully, provided conditions for growth as it created a need for these prospective teachers to know about planning. Through this activity, they could generate real and meaningful questions about planning for learning that I hoped might influence and enhance their approaches to planning in subsequent teaching experiences.

---

[2] In week 2 of the teacher education program, all students studying a Science method (i.e., Biology, Chemistry, Physics, General science) attend a two-day residential camp. The purpose of the camp is to introduce students to each other, the aims of the teacher education program and to do some teaching in small groups at a high school local to the camp site.

NEGOTIATING NEW ROLES

Resolving to teach differently, to withdraw (as far as I could) from a position of providing expert knowledge about teaching for Biology method students often left me unsure about what I might tell them that might be meaningful or useful for them as new teachers, or when it might be appropriate to tell them, if I chose to tell. I often wondered how I could help my students learn about teaching if I did not tell them what they needed to know? What did 'telling' mean, anyway? And, what sorts of conditions would support them to reflect and self-direct?

I knew that my role as a teacher educator was changing but I did not have a clear sense of what to do. One example of how I was wrestling with this aspect of my role is revealed in my attitude to summarising a class discussion. In week 3, I felt dissatisfied that I had summarised a student-led discussion, rather than encouraging students to do this for themselves. I felt as though in summarising for them that their ownership of their ideas was reduced, through imposing my meanings on their words. However, when I expressed my regret about my actions in the Open Journal one of them commented to me that in order to gain confidence as a teacher she needed me to reassure her that she was "on the right track". My summary of the discussion was one way of helping her feel reassured. Our comments from the Open Journal illustrate this point:

> At the end of the session I summed up the experience for everyone, told them what I saw/heard. This was the last thing they left the room with – me again! . . . I already know how to summarise a discussion – it is the preservice teachers who will most benefit from taking on that role.
> (Mandi: Open Journal Week 3)

> Student Feedback from Week 3:
> One of your comments last week was about your frustration that the last thing we left the class with was your summation of our learning instead of allowing us to do this. I agree that as preservice teachers we would gain valuable experience by leading discussions on what we thought. However, I also think that we need validation from you that we are on the right track . . . Confidence is a big factor in how we will succeed as teachers. If we are full of confidence we are more likely to be creative and daring in our approach. I'm not sure who else could give us confidence other than our tutors.

This student's feedback highlights an aspect of the tension between telling and growth as I experienced it. I believed that there were only two choices available to me: either I told these students about teaching, or I didn't. In this case I chose to tell them (i.e., I interpreted their experiences for them through summarized discussion) and then felt disappointed because I thought that my actions were in conflict with my beliefs about teaching (that they should learn to do this task for themselves). This student's response helped me understand (what in hindsight seems blatantly clear) that they needed me to tell them some things about teaching so that they could grow. It was the kind of knowledge and the context in which it was delivered that mattered.

This incident also highlights aspects of the second strand of the tension between telling and growth, that of acknowledging prospective teachers' concerns and challenging them to grow. Having the experience of summarising a discussion may have been helpful for them (challenging them to grow) however their over riding concerns were focused around feeling confident and reassured that they were "on the right track".

Thinking about prospective teachers' experiences of this tension led me to begin to reconsider my understanding of 'telling' and to make some small changes to my practice. Evidence that I was beginning to reconceptualise my approach and incorporate new ways of thinking about 'telling' into my practice is revealed in some comments to the class (weeks 15 and 17: video transcripts, below). While these examples might seem almost trivial, they do help to illustrate how the shift in my personal understanding was impacting my practice. I was beginning to feel more confident about what I could tell my students about teaching.

Week 15: Following a discussion about monitoring student understanding.

Mandi: Fantastic! You have brought up a variety of issues that as a teacher you're working your hardest to think about. 'How do I present things in a variety of ways? How do I know it's working?' It is complex; it is hard. And you're all doing a really good job trying to address these things.
(Video: Week 15)

Week 17: Following a discussion about ways of rewarding students who offer answers when they are not sure they are correct, and the value in this.

Mandi: Okay, so here's a great thing you've brought out. Teachers give this cue [about right answers.] If you've given the right answer they usually say, "Good, great!" If they haven't given the right answer they usually say, "Now, what do other people think?"

Sue: Or, "Anyone else?"

Mandi: Ok. What can a teacher do that doesn't give these cues, if you do genuinely want to explore students' opinions?
(Video: Week 17)

Just as it was important for prospective teachers to have opportunities to reflect and self-direct to stimulate their professional growth, so too I needed the chance to do this. The difficulty for both my students and for me, however, was that we were learning to do this together. I did not have a clear sense of how to do what I wanted to do, which created difficulties for knowing what actions I should (or should not) do to support others' learning about teaching.

EXPERIENCING THE TENSION OF TELLING
AND GROWTH AS A STUDENT IN MY CLASSES

In common with the experiences of other teacher educators who have introduced new ways of working into their classes (see for example, Carson, 1997; Grimmet, 1997) it was often difficult for students to know what I wanted them to learn about teaching, particularly if I did not explicitly tell them. Prospective teachers' expectations of learning to teach strongly influenced the way they perceived my approach. Those who expected *to be told* how to teach often found my approach unhelpful and frustrating, while those who expected *to learn* how to teach, found my approach more helpful (Loughran, 2007).

This point is illustrated in my interview with Bill (prospective teacher). Bill explained that his approach to learning and his view of himself as a learner led him to see that self-directed learning can be slow and difficult, and that sometimes it may not be worth the effort to learn this way if those things he needed to know about teaching could be told to him instead.

> Bill: . . . I'm an old fashioned sort of learner and I . . . like to know what's right and what's wrong . . . I like the sort of teachers who say to you don't do this but do these things, these 3 things are good . . . In anything I'm learning, I like to know. That's like the dictatorial style of teaching. These are the things you should and shouldn't do, these are the types of things you should try to emphasise and away you go. Sometimes when you are just thrown in at the deep end type learning, it can be a long process, before you actually might come to grips with something that you've been doing wrong for a long time, because no one has actually said to you, "When you stand out the front why don't you just smile, just once, it'll change the whole feel of the thing?". . . No one actually says what you're doing wrong specifically . . . and I think a lot of us need to be 'brushed up' individually.
> (Bill Interview 1: 260–262)

Bill's use of words such as *"old fashioned"*, *"dictatorial"*, *"what's right and wrong"* indicate a view of knowledge about teaching as externally produced, certain and transmitted unproblematically from sender to receiver. Bill's (short term) desire to be told 'what works' was in conflict with my (long term) goal of helping him to move towards personally oriented growth (Mason, 2002). Perhaps unsurprisingly then, learning about teaching in my classes became a very frustrating experience for Bill. (Further examples of this are explored in other tensions.)

Lisa took a different view of my approach. Lisa's view of knowledge was such that it was personally constructed and resulting from an interaction between her experiences and her thinking about these experiences. In order to develop as a teacher she needed to make sense of her own ideas about how teaching worked for her. In our second interview Lisa identified that her learning about teaching had not come about as a result of my telling her what she needed to know, but from the questions that she had been encouraged to ask of herself and others.

Lisa: I reckon you've influenced me mainly because you've helped me think about my teaching. Like you haven't helped me because you've said, "Oh this is how you do it". You helped me think about how I want to teach and what I wanted to get out of teaching, so you've helped me think about that just on my own and then you've also allowed all of us to ask you questions about why you do things and then we can think about whether it's worth doing that . . . and that's really helpful . . .

Mandi: Because I have explicitly invited people to ask questions or you've picked up that it's ok to ask questions?

Lisa: Oh, I think it's the latter. Even though you have invited us, yeah, you've invited us, but not everyone took up the invitation.
         (Lisa Interview 2: 234–251)

Bill's and Lisa's views represent opposing expectations of learning to teach. Bill sees the authority of the teacher educator as ideally positioned to be able to correct teaching mistakes and in this way help him learn about teaching. Lisa, on the other hand seems to find looking into experience (hers, mine, others) a helpful way for her to learn about teaching. It is not surprising then, that Lisa responded to my approach with more enthusiasm than Bill.

My struggles to enact a pedagogy of teacher education with my students that genuinely created opportunities for them to reflect and self-direct also meant that from time to time, students were left unsure about my purposes for their learning. Lisa described my approach as 'subtle' and that as a consequence, she wished for some more clear direction from me about what she was expected to learn. At the time I found her comment puzzling, because even though I had resolved not to tell prospective teachers about how to teach, I did think I was being explicit about what was the broad purpose of each session. In fact, I did not know what else I could do to make my purposes for their learning more explicit.

Lisa: . . . sometimes . . . you are pretty subtle. And I think I wish she would just tell us what she wanted, even though I think that you think that you are being really explicit. I reckon I'm like that as well. I have got it in my head that I know what I want them to do but they have got not much of an idea . . .
         (Lisa Interview 2: 120–129)

Lisa referred to my approach as 'subtle' on several further occasions (Lisa e-mails: April 02, 03, June 27).

Interestingly, a student from the previous year's Biology methods class, whom I interviewed in a pilot study for this research, also raised the issue of needing to have a clearer sense of my agenda for her learning. Anne spoke about her experience of Biology methods classes:

> Anne: [It] felt like we were on a journey but we didn't know where we were going. I didn't know where I was going but I knew that you were taking us somewhere . . . Do you think it would have been better if you had told us what you were doing, or would that have interfered with you not wanting us to know what you thought was good practice so we could work it out for ourselves?
>
> Mandi: What would have helped you in terms of explaining the destination?
>
> Anne: . . . Not by saying this is the destination, but I hope you will question how we'll get there. My goal is to get you to question, to open your mind to other possibilities. That's not [the same as] you giving them the answer to what it is. Then at least people know that's what she [Mandi] wants us to do. I think that's helpful . . .
>   (Anne, Pilot Interview: 2000)

Anne's message to me is very similar to Lisa's – prospective teachers needed to know more about my purposes in constructing the Biology methods course in the way that I did. Anne acknowledged my reluctance to tell prospective teachers how they should teach, but suggested there was still useful information that she (and other students) could be told to help them make better sense of my intentions for their learning. I did feel that I had attempted to address this aspect of my practice in the following year (i.e., 2001) but Lisa's comments suggest to me that I did not do so in a way that was satisfying for her (and possibly other students, also.) This further highlights my struggle with 'telling'. An ongoing challenge for me within this tension is in coming to terms with the notion that there is a difference between telling prospective teachers about teaching and giving them sufficient information about my intentions to make clear to them what they are expected to learn.

Making the Tacit Explicit in My Practice as a Way of Gaining Insight into My Purposes

One way in which I tried to facilitate prospective teachers' understanding of my intentions for their learning was by making explicit my pedagogical reasoning while I was teaching. This aspect of my practice is discussed more fully in other chapter tensions. However, it was an approach that I used with the purpose of helping students to gain some insights into what was influencing my pedagogical choices. In this way, I hoped that my students might be prompted to consider teaching as a process of decision making (as opposed to a series of routines) and, as a result, to begin to think more deeply about their own pedagogical decision making. I anticipated that through the process of 'thinking aloud' about my practice that my purposes for their learning would be made clear. In hindsight I recognise that there is an important difference between providing access to an experienced teacher's thinking and prospective teachers knowing why it is being offered or how my thinking is linked to what I expect them to learn about their own teaching.

One reason why prospective teachers may have found it difficult to understand my intentions for their learning might have been because, in the process of making my thinking explicit, I was paying more intention to *my* needs and concerns, about how I should behave as their teacher educator, (i.e., not telling), and was therefore less sensitive to the particular conditions that might enable them to hear what I was saying and begin to grow. This idea links to the second strand of the tension between telling and growth; that of acknowledging prospective teachers' needs and concerns and challenging them to grow beyond their immediate preoccupations. Other teacher educators have highlighted this tension within their practice. For example, Nicol (1997b), a Mathematics teacher educator, found that as a consequence of studying her interactions with her students that she became more sensitively aware of the balance "Between accomplishing . . . [her] own teaching goals and experiencing teaching through prospective teachers' eyes" (p. 112). As a consequence of her study, Nicol began to "reframe" (Schön, 1987) her understanding of her practice as she learnt to recognise differences between when she was introducing her own agenda and when she was responding to prospective teachers' particular needs. Awareness of this aspect is important in shaping what it means to challenge expectations of learning to teach in order to develop a deep understanding of that which will influence one's own practice.

Difficulties of Looking Beyond My Own Needs in Helping Students to Grow

The notion that I was responding to my own needs for students' learning rather than listening to what were students' needs for their learning became apparent in my efforts to challenge students' thinking about their pedagogy during their peer teaching sessions. My view was that by asking probing questions about the teaching approach chosen and about learners' responses to the teaching in the debrief following each peer teaching session (and encouraging prospective teachers to do the same), that this would create opportunities to learn and grow through collaborative critical evaluation of the teaching and learning experienced. However, my intentions met with limited success in practice. One difficulty was that a number of students found the experience of debriefing too confronting. Several students reported that they felt uncomfortable critiquing each other's teaching, so that instead of opening up conversations about practice and creating new possibilities for action, prospective teachers were more inclined to pose non threatening questions (e.g., "why did you choose this approach?"), or to defend and rationalise their teaching approach when questioned by me or their peers in ways that may have been perceived by them as too challenging.

Maybe had I stopped to consider more closely their needs and concerns then the approach to debriefing that I chose may have been different. At the same time however, I *did* want to encourage them to think differently about their practice. This strand of the tension between telling and growth becomes real when the teacher educator sees the conditions for growth as ones that challenge the existing views of the prospective teacher and must therefore ask: when is challenge productive, and when is it destructive? My focus was on helping prospective teachers analyse their teaching experiences for productive professional growth, but some clearly felt that my actions were detrimental to their learning about teaching.

WHERE VIEWS COLLIDE

A clear difficulty for me associated with helping students to grow beyond their immediate preoccupations lay in recognising what conditions for growth might be suitable for different students. Because I was strongly aware of my own concerns to reduce the supply of propositional knowledge about teaching, I tended to be highly sensitive to this orientation in others and, consequently far less supportive towards them. In fact, I often took a more confrontational approach in situations where prospective teachers were operating from a transmission model, rather than acknowledging and supporting them so that real possibilities for their pedagogical change might be created.

On the other hand, those who showed that they were struggling against 'telling', or who were experimenting with new approaches to practice, were more likely to receive my support and encouragement. Rather than considering what might be suitable approaches to dealing with individual students on the basis of their individual needs and concerns, I implicitly worked from my own. Interestingly then, it was Lisa's confidence, developed from supporting, rather than challenging her, that led her to push the boundaries of her own learning and to challenge herself to extend her understanding of practice. Perhaps had I recognized the benefit of supporting and trusting *all* of my students, even those whose models of teaching were different to mine, then more opportunities for extending their thinking and exploring alternative models of pedagogy may have been created. Belenky, Clinchy, Goldberger and Tarule (1986) acknowledge this point in highlighting the role of trust and being sensitive to the particular needs of individual students even though as a person or critic one might not agree with their views or approaches.

> To trust means not just to tolerate a variety of view points, acting as an impartial referee, assuring equal air time to all. It means to try to connect, to enter into each student's perspective. (p. 227)

Negotiating the tension between telling and growth was a task that I was attempting to understand and manage at the same time that the prospective teachers in my classes were attempting to understand and manage this for themselves. I needed to acknowledge and work from this dual perspective in order to better help them learn to negotiate this tension for themselves, and allow them to grow; supported by my trust.

Other, external influences such as the school context, also shape prospective teachers' expectations of their learning about teaching. Such influences further contribute to the tension between telling and growth.

The Telling Model and the Influence of the School/Curriculum Context

Prospective teachers' beliefs about teaching Biology are strongly influenced by their own experiences of learning Biology. At the senior levels of secondary schooling in particular, there is considerable pressure on teachers to 'teach to the exam', so it is not difficult to see how getting through the curriculum can drive a teacher's agenda. Transmissive approaches to learning are frequently rationalized by teachers (and students) as an unfortunate necessity, with understanding a desirable, but often impractical goal. For example, Trumbull (1999) conducted a longitudinal study of several new Biology teachers as they moved from university to their first paid employment as teachers in schools. Trumbull explored the ways in which these new teachers began to recognize and grapple with contradictions between teaching for understanding and teaching to pass examinations. Such a view of teaching Biology impacts on prospective teachers' expectations of their teacher preparation program as they may see themselves confronting an 'either–or' situation of whether they are learning to teach Biology to help students pass exams or learning to teach Biology for understanding. Practicum experiences, including expectations of supervising teachers and school students often serve to reinforce notions of exam passing as a teacher's main responsibility which, in turn, influences the model of teaching that prospective teachers choose to employ.

In hindsight, I realize that I did little to help my students better understand the nature of the situation they faced as new senior secondary Biology teachers. While we talked a great deal about the pressure they felt to teach in particular ways (i.e., 'teach to the test', 'tell students what they need to know') I didn't explicitly talk with them about the nature of this problem or how they might perceive their choices for teaching as limited to an 'either–or' scenario.

As an experienced Biology teacher I knew that there were many instances when I 'forward feed' high school students particular content knowledge. Confidence in knowing the structure of the curriculum, including which concepts require more time and which less, what is a difficult concept or a foundational idea compared to that which builds on foundation knowledge makes a huge difference in successfully negotiating these complexities to make an informed decision. This situation raises interesting parallels between prospective teachers' perceptions of their new teaching role and my own.

Inexperience in our roles (as prospective teacher and teacher educator) tended to produce a view that our choice of teaching approach was limited to an 'either–or' situation. Gaining confidence and experience (such as I have in teaching high school Biology) enables the possibility of rejecting such a restricted view and to embrace ideas of 'both–and'. In other words, experience enables the teacher to begin to see situations as more complex than that represented by a binary view and at the same time, can help the teacher recognise that there are some things that cannot be told. Choosing what to tell and when to tell it is what matters most.

SUMMARY: WHAT DID I LEARN FROM EXAMINING THIS TENSION WITHIN MY PRACTICE?

The desire to tell prospective teachers about teaching is strong. The seductive nature of telling often leads teacher educators to overestimate the extent and impact of what can usefully be told to those learning to teach and underestimate what prospective teachers can learn for themselves (Northfield & Gunstone, 1997). Lisa highlighted this very problem when she described the effect on her of a particular teacher educator who did a lot of 'telling' and neglected the students' experiences as opportunities for learning. From this experience, Lisa considered the implications for her own teaching:

> Date: Tues, 6 March 2001
> From: Lisa
> To: amanda.berry@education.monash.edu.au
>
> . . . I guess she[lecturer] is very concerned with getting her own story across that she has lost the idea that we can learn from our own experiences and from each others. This is an important lesson for me because I feel sometimes that I have so much to share and it will be hard for me to let my classes learn for themselves. I will have to buy a stapler and staple my lips together. Maybe I could put a stapler on my desk to remind me to SHUT UP! . . .

Lisa pinpoints an aspect of telling, that it is far too easy to tell prospective teachers what they need to know and to offer vicarious experiences of teaching and learning through recounting one's own memories of teaching, and ignore the value in allowing prospective teachers to grow through offering them experiences that might teach them about learning in such a way as to develop their own understanding of these ideas.

In my approach to teaching Biology methods, I chose to deliberately withdraw the authority of my experience (Munby & Russell, 1994) as a 'teller' and to create experiences that encouraged prospective teachers in my classes to construct their own, personally meaningful, knowledge about teaching. In so doing, I did not anticipate the difficulties that I came to face when my students did not construct knowledge about teaching in ways that I anticipated. The contradiction between my intentions and how my students experienced my intentions in Biology methods classes, created the tension between telling and growth described in this chapter.

The tension between telling and growth is an example of finding a balance between the applicability and value of different forms of knowledge. Propositional knowledge that applies generally to many different situations is frequently formulated in abstract terms and traditionally offered to prospective teachers in a "teaching as telling, showing, guided practice approach" (Myers, 2002, p. 131). On the other hand, the development of knowledge through experience involves developing a sensitivity to situations and a concentration on decision-making about what might be helpful for teacher educators to highlight (or not) for their students in a given situation, and/or how to highlight a particular issue in a given situation.

Teacher educators who wish to challenge the traditional model of 'teaching as telling' are therefore confronted by a real tension. It may be clear what prospective teachers "need to know", but this is very different from them knowing how to act. Hence the teacher educator struggles between informing (delivering the propositional knowledge) and creating opportunities to reflect and self-direct (making experiences about the issues personally meaningful). This tension is also exacerbated by moderating between acknowledging prospective teachers' needs and concerns and challenging them to grow.

Chapter Six

CONFIDENCE AND UNCERTAINTY

Confidence (n) firm trust. A feeling of reliance or certainty. A sense of self reliance; boldness.

Uncertainty (n) subject to doubt; not fully confident or assured, tentative.

> Old ways of thinking no longer make sense but new ones have not yet gelled to take their place, leaving one dangling in the throes of uncertainty. Yet, this uncertainty is the hallmark for transformation and the emergence of new possibilities . . . (Larrivee, 2000, p. 304)

INTRODUCTION

Similar to the experiences of other teacher educators who have moved away from traditional models of teaching about teaching to explore new ways of working with prospective teachers (see for example Emert, 1996; White, 2002), I began to experience feelings of self-doubt and uncertainty about how to proceed in my new role. I knew that I wanted to change my approach to practice and, to some extent, I possessed the formal knowledge to support my reasoning, yet I did not have the personal, experiential knowledge to carry out the role in the manner that I envisioned. At the same time, I recognized that prospective teachers beginning their preservice preparation also experienced feelings of uncertainty and self-doubt about their competence as new teachers. And, as a consequence, prospective teachers usually (quite naturally) look to their teacher educators to help them develop certainty and confidence in learning 'the right ways to teach'. This situation yields the source of the tension between *confidence and uncertainty* as I experienced it in my teacher educator role: How could I build prospective teachers' confidence to grow as new teachers, when I was questioning the nature of my teaching about teaching?

Two separate strands comprise the tension between confidence and uncertainty for me. One strand relates to exposing my vulnerability as a teacher educator as I seek to develop my teaching in a manner consistent with my beliefs, while at the same time attempting to maintain students' confidence in me as their leader. The other, relates to the deliberate choice I made to expose the uncertainties behind my teaching as a

prompt for prospective teachers to begin to view teaching as an uncertain and
problematic enterprise, while at the same time helping them to feel sufficiently confident that they could successfully develop as new teachers. Each of these strands is
now elaborated. I illustrate them through my experiences and subsequent growing
understanding of helping prospective teachers learn about teaching while at the same
time that I was questioning the nature of my teaching about teaching.

CONFIDENCE MATTERS

For prospective teachers and teacher educators alike, confidence matters. An important
goal of my teaching is to help students develop as confident new members of the teaching profession. However, I have learnt that my understanding of confidence differs considerably from that of my students. To me, confidence means being open to exploring
new possibilities for one's teaching, willingness to listen to the ideas of others and preparedness to make changes to teaching in order to facilitate more meaningful learning.
These ideas are embodied in Dewey's (1933) personal attitudes of openmindedness,
wholeheartedness, and responsibility. In my approach to preservice Biology teaching, I
sought to represent these attitudes as a form of confidence; a boldness to explore with
prospective teachers the relationship between my teaching and their learning.

This notion of confidence carries a somewhat different meaning when considered
from the perspective of a beginning teacher. Many see teaching as an uncomplicated act
of telling students what to learn – a consequence of years of uncritical observation of
their own teachers at work (Britzman, 1991; Lortie, 1975; Pajares, 1992). Consequently,
beginning teachers may enter pre-service programs with an expectation that they can be
told how to teach and therefore appear to be in search of a recipe for teaching. This
means for them, that confidence comes in the form of a recipe that may well comprise a
set of practical teaching strategies that will be seen to ensure their success in the classroom. Prospective teachers may therefore be critical of their teacher preparation program if this does not occur (Britzman, 1986): a lesson that I quickly learnt.

In our first interview, Andy described his expectations of the teacher preparation
program. Andy imagined that he would be provided with "information" about teaching, although he said that he had not really thought a lot in advance about what to
expect from his studies (a common response from prospective teachers).

Mandi: So how do you feel about Biol method compared to what you hoped or
         anticipated Biol method might be like?

Andy: I didn't think about it much . . . I had some idea because my girlfriend
         did it last year . . . I suppose an idea I had was that it would have been
         more structured in that you'd roll up to class and you'd [lecturer]
         stand up the front and say this is what we're doing today and you'd be
         given handouts, you'd be given information, you'd download it to
         your brain and you'd take notes or whatever and you'd walk off. Sort
         of old fashioned teaching and learning . . . These are the methods that

you use; you can have option 1 to 10. This is how it's done. These are
the techniques . . . the tried and tested ways . . .
(Andy Interview 1: 249–260)

Another prospective Biology teacher, Bill, compared his experiences of learning
to teach in Biology methods classes with his experiences of learning to teach in other
subjects in the teacher education program. He expressed disappointment that he was
not given the kind of preparation in Biology methods that other subjects offered.
This led him, (and he reported similar feelings from other Biology methods students), to feel under prepared (i.e., less confident) for the teaching role.

Bill: I mean there's a general feeling, and I think I speak on behalf of quite
         a few people, that your Biology is quite different to the other . . .
         classes as to what preparation is being given us . . . to get us ready for
         it [teaching] and the answer's not much . . . I see that you are dealing
         with us on a more cerebral level than perhaps a lot of the other method
         teachers. Do you know what I'm saying?

Mandi: Can you tell me about what you mean?

Bill: Well, I think you're trying to, you know, get us to do our own thinking,
         but the others aren't, and so we feel like we're being neglected a bit [in
         Biology]. That's probably a bad word, neglected's not the right word
         but we're, we don't feel well armed in Biol . . .
         (Bill Interview 1: 111–119)

In reporting this exchange, I feel pleased that my actions in Biology methods
were interpreted by Bill as encouraging self-reliance (since encouraging prospective
teachers' independence was an important goal of my teaching), yet also disappointed
that what I was doing was not perceived as helpful for Bill and some of his peers.
Bill then told me in order to build his confidence and competence as a Biology
teacher it would be helpful to do as (some) other method lecturers had done and
work towards more concrete outcomes, for example building a resource bank of
activities that could be used with school students. My discomfort in this situation –
how could I help Bill feel more confident about the usefulness of what I was offering
for his learning about teaching Biology, when I was not sure myself – illustrates the
tension of confidence and uncertainty as it played out in my practice.

Bill: I think that what people have said to me was that – and this could be
         wrong – that they feel like they're not getting any methods of teaching
         Biology under their belt . . . like in [other method] we spend most of the
         time digging up resources and activities for each other and we're collat-
         ing this pile of activities and games and interesting things. And we're
         just piling it all up for the VCE[1] study guide . . . in order that we all will

---

[1] Victorian Certificate of Education: state mandated curriculum for final years of secondary schooling.

get a copy of everything that everybody's collected. You know, a
resource that you'll be able to use every day of the year. But Biol's not.
When I look at my folder there isn't much that I can see. Now that's up
to me I can see that, but I just thought while we were here, maybe we
could do a bit more of that.
(Bill Interview 1: 157–160)

My approach to planning and teaching Biology methods was to look beyond this
view of teaching as the stockpiling of activities. I wanted to challenge the notion of
teaching as an uncomplicated act of following particular 'tried and tested' routines.
My goal was for my students to develop their own personally meaningful approaches
to teaching that were also congruent with supporting their students' effective learning of biology concepts. (My experience suggests that ordinarily, most 'activities' do
not support this kind of learning.) However, in so doing, I neglected to acknowledge
the needs of the prospective teachers I was working with, such as Bill.

For many of my students, their sense of confidence was dependent on having
'ready made' resources at hand, to help them feel prepared. This is a critical aspect
of the dilemma (and hence the source of the tension) faced by many teacher educators who choose to risk the uncharted territory of new approaches to practice that
challenge traditional expectations held by prospective teachers and Faculties of
Education. When the implicit message from prospective teachers is for the teacher
educator to supply the knowledge and resources for teaching, and prospective
teachers' predominant experiences of learning about teaching seem to reinforce this
view, then it is enormously difficult to avoid feelings of self-doubt and inadequacy
as a teacher educator if one's practices do not conform to these expectations.
Feiman-Nemser and Remillard (1996, p. 76) draw on the work of Floden and
Buchmann to note this very point: "Just as uncertainty challenges teachers, so
preparing teachers for uncertainty challenges teacher educators". (At the same time,
it is a difficult task for prospective teachers to suspend their immediate needs for
'certain' knowledge and to take a 'leap of faith' to trust a teacher educator who
chooses to work differently – a point that will be taken up later in this chapter.)

BUILDING (AND LOSING) CONFIDENCE

The development of confidence in one's competence is a tricky thing because it does
not occur as a 'lock-step' progression. Confidence can be quickly and easily undermined in the process of learning to teach; from a comment or experience. The
'undermining' effect of this situation on one's confidence is amplified when the view
of teaching held is that there is 'one right way'.

Bill: . . . Just when I feel like I've got a grip on what this teaching thing is all
         about, because it was all a bit vague at the start, as all things are when you
         first try them, and I was just about sort of ok, I can get my hands around
         this, some criticisms had come in from wherever and then I felt insecure

again, because I thought I had the hang of this . . . [but] I'm not doing it the right way. I feel I'm using my personality, trying to get what I've read and what I know across to them [students] in a way that's going to be reasonably easily explained, varying the stimulus, you know, activities, bit of talk, bit of humour . . . overheads, all that sort of stuff, and then, you find out that you've missed something . . . It's like trying to swim in the dark to the end of the pool and you think you're there and you sort of turn the light on and you find out you're only a third of the way there. Am I ever going to get to the end? That's how I sort of felt a bit . . .
(Bill Interview 1: 90–94)

Bill identified that feeling confident in his teaching abilities was important to him. He wanted to "get the hang of" teaching. However, his confidence was undermined after he received some negative feedback (although it was not clear where this came from) about his teaching that led him to feel insecure and frustrated. Despite his efforts to incorporate the techniques that he had learnt to teach "the right way" (i.e., straightforward explanations, variety in presentation, and a personally engaging style), he discovered that there was more to teaching than these elements alone, and more than he could actually see at this stage of his development.

Later, when Bill was faced with a situation in which I revealed my professional uncertainty about how to proceed during a methods session (Week 4), it did not serve as a stimulus for him to reflect on his own approach, as I had intended. On the contrary, he was frustrated and puzzled (as to why I would want to share my uncertainty) about what I was doing because it was not helpful for his development as a teacher. My actions unsettled him because Bill experienced a strong need to feel sure about his competence as a new teacher.

A different element contributing to prospective teachers' need for confidence in their development comes from the subject matter of Biology, particularly the knowledge status of Biology as a senior secondary school subject.

Subject Matter Confidence

My experiences of teaching Biology methods have led me to recognize that prospective Biology teachers often begin their preservice education feeling that they have insufficient content knowledge about Biology to be successful teachers of senior Biology. Because their view of teaching Biology is commonly associated with effectively supplying large amounts of information to their students (with effectiveness measured by their students' success in external testing procedures), they believe that they must know all of this content knowledge well, in order to supply it. Prospective teachers (and new Biology teachers) usually feel an overriding responsibility to ensure that their students can leave their classrooms with the correct and appropriate scientific concepts (Hewson, Kerby & Cook, 1995). This is particularly the case for senior level classes, where "student teachers feel constrained to 'honor' the content more and their students' thinking less" (Hewson, Tabachnick, Zeichner & Lemberger, 1999, p. 381).

Such a view of Biology teaching is often reinforced by supervising teachers during the practicum, who criticise preservice teachers for their lack of content knowledge appropriate to the Biology curriculum. This in turn, impacts enormously on how they measure their competence as beginning teachers, what they want to learn about teaching Biology, and how they respond to teacher educator messages that do not conform to these ways of thinking about how to teach Biology. In hindsight, it was not surprising that the approach to learning about Biology teaching that I took, with its emphasis on exploration and questioning of traditional transmission approaches and consideration of biological knowledge as uncertain rather than 'timeless truths', mattered much less to many of my students than did their need to feel confident about their content knowledge and effective ways to impart it. Our purposes were at odds, with their most pressing concerns for confidence and certainty, reinforced as legitimate by most of what they saw around them within the teacher education program and in schools. The effect of this situation (as illustrated by Bill's comments, earlier) was that my approach could readily be dismissed as irrelevant or unhelpful.

Ironically, uncertainty characterizes modern Biology, as science research challenges long held understandings of Biological concepts (e.g., inheritance, cell structure and function) and offers incomplete or contradictory explanations of new knowledge in Biology. Fisher and Lipson (1986) noted that:

> ... teachers in rapidly changing fields such as the biological sciences become almost as accustomed to uncertainty and error as their students. They must constantly accommodate as the accepted 'facts' and paradigms of field shift around them ... Teachers need to impart the tentative and imperfect nature of their judgments to students so that students can maintain their morale and energy as they learn. (p. 789)

For many preservice Biology (and Science) teachers, a sense of confidence to act as a teacher is, at least initially, derived from knowledge of the Biology content material since it gives these new teachers something to 'teach', even though they may not yet know how to teach it effectively. Accepting a view of Biology knowledge as uncertain is therefore an unsettling proposition for many since it makes problematic what Biology to teach at the same time as they are struggling to know how to teach Biology.

In hindsight, I was unaware of the possible impact of my approach on disturbing prospective teachers' confidence as 'knowers' of Biology. My own need to pursue a particular approach in my teaching about teaching, i.e., to help prospective teachers recognize and welcome uncertainty as a characteristic of teaching and of Biology, was paramount and blinded me to their need for confidence in order that they might consider trying new ways of thinking about, and working with, unfamiliar practices.

Teacher Educators Struggle with a Need for Certainty too

It is not only prospective teachers who desire a sense of certainty in their work. Teacher educators can also feel the need to know that what they are doing is 'right'. This is particularly difficult when their teaching takes them away from the familiar path. Emert (1996), a teacher writing about the development of her own practice, identified her feelings of resistance to change because of the lack of certainty that new practices offered. Emert chose to live this contradiction by continuing to enact certain approaches to practice, even as she recognised that these were not effective for her students' learning.

> Many times I felt I did not know what I needed to be an effective teacher. I was frustrated with what I perceived as a lack of knowledge and also a lack of trust in the knowledge that I did have. This insecurity caused me to avoid trying some new practices. Often I would stall and extend certain experiences that I felt comfortable with. At other times I would revert back to old activities that had little relevance for my students. This stalling was done in fear of change. I was afraid I was not doing things "right" or the explorations were heading in the "right" direction. (Emert, 1996, p. 263)

Emert constructed a teaching style that allowed her to avoid, to some extent, the feelings of discomfort that came from her uncertainty, and that impeded her ability to step out into the unknown. Reading Emert's narrative helped me to recognise more about myself. While I professed a belief that uncertainty and doubt were central to the development of mine and my students' practice, my own needs for certainty and a desire to avoid insecurity led to feelings of confusion about how I could enact my beliefs. There is a considerable amount of risk involved in exploring practices that have unknown outcomes (for prospective teachers and teacher educators alike) particularly when the implicit message from prospective teachers is for the teacher educator to be seen as a competent expert. Feelings of incompetence experienced by teacher educators can be communicated to students, causing unnecessary levels of mistrust or doubt (Brookfield, 1995) – a point that will be taken up later in this chapter.

I struggled to reconcile the idea that if I subscribed to a view of knowledge that embraced uncertainty and ambiguity, then I could not expect to know a 'right way' to act or to believe that a 'right way' even existed. (The paradox of this situation was lost on me at the time: I was encouraging prospective teachers to relinquish their ideas about certainty, yet I struggled to do the same myself. Somehow for me, the risks seemed different.)

An example from my practice that illustrates this very point comes from the Biology methods class in Week 3. I had organized the class to work in groups then, as they set to work, I observed their group interactions. The journal extract below is my response to watching the videotape of the class group work task in action:

> I have to accept that this is normal [behaviour], even though I feel disappointed [about non task specific behaviour as students begin to work in groups]. It's probably about establishing social organization or something. But how do I bring this to their attention for discussion as something that will happen in their classes, also? Should I bother bringing it to their attention? How do I know? (Personal Journal, Week 3)

I assumed that because I had set a task for the class that they would immediately begin it when they moved into their groups. The fact that most groups did not begin to work on the task immediately, surprised me and unearthed an assumption of mine about my influence as a teacher and my expectations about how group work operates. At the time, I was unsure about how and when, or even whether to bring this issue up for discussion with the students. Although it is a relatively minor example, it serves to illustrate the tension of uncertainty in my pedagogical approach. Consideration of the issues associated with knowing what aspects of my uncertainty to reveal to students links to the second strand of the tension between confidence and uncertainty. This strand played out through my efforts to make explicit the complexities and messiness of teaching, while at the same time attempting to help prospective teachers feel sufficiently self-confident that they could make progress in their learning about teaching.

MAKING EXPLICIT THE COMPLEXITIES OF TEACHING

The idea of teacher educators deliberately making explicit the complexities and messiness within their own teaching departs from the regular practices of teacher education. Teacher educators do not usually reveal to their students the problematic nature of their work. The motivations of the teacher educator (or school based supervising teacher) in devising the lesson, the alternatives that she considered in planning or implementing the lesson, or the (sometimes problematic) decision making that she faces from moment to moment within the lesson are rarely made apparent to prospective teachers. As a consequence, prospective teachers commonly see teaching enacted, within schools and education faculties, as little more than a set of smoothly executed skills and routines. Consequently, the pedagogical significance of what is being observed can make teaching appear deceptively simple.

Explicit Modeling as an Approach to Learning about Practice

I believe that teacher education should provide opportunities for pre-service teachers to 'see into' teaching practice in ways that stimulate examination of existing perceptions and encourage consideration of alternative frames of reference. In this way, they may be motivated to think about new and more complex understandings of teaching and learning. It is in the experience of teacher educators opening up their teaching practice to their students so as to make explicit their thinking about the uncertainties, dilemmas, questions and contradictions that they face in their own teaching, that new possibilities for practice can emerge.

In my approach to teaching about Biology teaching, I wanted prospective teachers to become more aware of their processes of pedagogical decision-making, so that they might be more thoughtful about the pedagogical choices they made. I chose to work towards this goal by explicitly modeling my own decision-making processes for them.

Explicit modeling operates concurrently at two levels. At one level, explicit modeling is about 'practising what I preach'. This means, modeling the use of engaging and innovative teaching procedures for prospective teachers rather than 'delivering' information about such practices through the traditional (and often expected) transmission approach. At another level, it means offering students access to the pedagogical reasoning, feelings, thoughts and actions that accompanied my practice across a range of teaching and learning experiences. I used two main ways to make my thinking about my practice explicit to prospective teachers in the Biology methods class. First, in class, I used a process of "thinking out loud" (Loughran, 1996) about my teaching as it was happening. This involved, for example, selecting moments from time to time during the class to stop, highlight and examine with the students, particular uncertainties I was facing or consequences of teaching decisions I had taken. Second, I published an Open Journal of my thinking about my practice to a Webpage linked to the Biology methods homepage within the Faculty of Education, at Monash University. My Open Journal documented my purposes for each method session and reasons for these purposes, reflections on what I felt and observed as I experienced teaching the class, and ways in which the events of the session helped to inform my thinking about my pedagogical purposes or practices. While both of these approaches serve the same broad purpose, i.e., making my thinking about my practice explicit to others, there were important differences between them.

Belenky et al. (1986) identify the importance of teachers revealing the thinking behind their actions for their students:

> [T]he teacher . . . takes few risks . . . he [sic] composes his thoughts in private. The students are permitted to see the product of his thinking, but the process of gestation is hidden from view. The lecture appears as if by magic. The teacher asks the students to take risks he is unwilling – although presumably more able – to take himself . . . So long as teachers hide the imperfect processes of their thinking, allowing the students to glimpse only their polished products, students will remain convinced that only Einstein-or a professor-could think up a theory. (p. 215)

Revealing "the imperfect processes of [my] thinking" is what I am calling the second level of explicit modeling. But fostering a climate that supports sharing of confusions is not easy (Trumbull, 1999). Persevering with explicit modeling carries with it a certain amount of tension and uncertainty. Making a choice about what to make explicit both in my talking about practice during classes and in my other communications with prospective teachers was a constant dilemma. I had to choose carefully what I held up for public examination that would be useful and accessible for these prospective teachers. In hindsight, I do not think I really recognized how the different "scripts" (White, 1988) that we carried for teaching affected their perceptions of what I said or wrote. I wanted to convince the Biology methods students that it was okay to be unsure of one's own practice; to see that teaching is a problematic venture.

Journaling as a Means of Opening up Practice
for Scrutiny and Dialogue

My decision to keep an Open Journal was in part motivated by some of the difficulties
associated with what aspects of my thinking to make explicit to prospective teachers (it
was often difficult for me to decide such matters 'on the spot' in methods classes) and to
encourage prospective teachers to re-think their experiences of a session, perhaps even
engage in a conversation (electronic or otherwise) about practice with me and/or their
peers. Writing a post session reflection in the quiet of my office gave me time to examine my actions, thoughts and feelings from a class and so decide what might be most
useful to bring to these students' attention for their learning about teaching. I used the
Open Journal to articulate some of the uncertainties I experienced in planning classes,
so that students might see that teaching is based on making decisions about what and
how to teach, and that these decisions are moderated by the learners and the context.

For example, in Week 1 of Biology methods, I made the following entry in my
Open Journal:

(Entry written prior to method session)
I have been in two minds about how to proceed this week. First, it would be
helpful to continue exploring the idea of living and non living and to look at how
possibilities from last week's session might be followed up in the classroom. In
other words, if the teacher has generated confusion and questions (as I did) then
how might the teacher move on from that point to help students start to organise
their knowledge about that idea? However, on the other hand I feel like it is a bit
soon to proceed with resolution of these ideas just yet. Presenting the samples of
materials and encouraging question generation is a good teaching approach in
itself because it aims at exploring students' knowledge rather than the teacher's
knowledge, and that is a message that I wish to encourage in thinking about
teaching. I'm also hoping it will provide a model for how the group will go
about their 'interview with a student' assignment[2]. It is important to gently
probe someone's thinking rather than closing in on one right answer, if you want
to get an accurate idea about that person's thinking. . . . I am aware that I am
trying to do this, but how clear is my intention to the group?? I wonder whether
anyone has followed up on the idea of living or nonliving?? How much does it
matter amongst other competing concerns??
(Entry written after the session)
It was fantastic today to have spontaneous discussion about whether people had
gone off and talked with others outside uni about the meanings of living and non
living. I was so pleased that this discussion wasn't initiated by me and it was
something that provoked a response in a few people e.g., one person consulting a
text and another discussing with a friend. It would have been worth pursuing, but

---

[2] Assignment in which prospective Biology teachers interviewed middle school students about their
understandings of a biological concept.

I didn't, to find out what motivated some to continue to think about the issue while others forgot it as soon as they left the room.
(Open Journal, Week 1)

In the extract (above) I publicly wrestled with the issue of how to deal with the confusion that I had purposefully generated in the previous week from asking students to identify a selection of objects as 'living' or not. I could help students resolve some of their ideas through further discussion and questioning, or I could use the experience as a way of modeling an approach to eliciting learners' different ideas, and not worry about taking it any further. Either way enabled me to model a constructivist approach to learning – although different aspects. In the end, it was the enthusiasm of a few students who sought to resolve their own confusion that led to a whole group discussion about the meaning of the word 'living'. (Interestingly, I did not use this experience as a means of helping students to consider the process of learning that they had engaged in.)

From time to time in class, I drew students' attention to issues that I had addressed in the Open Journal, hoping that this might encourage further engagement with these ideas. A small number of students posted electronic responses to my Open Journal entries. (Typically students chose an aspect of a session to discuss, usually to offer an alternative perspective of an episode we had experienced, or to voice a concern about some aspect of their own or others' learning in the class.)

I also surveyed the class to find out their views on the purpose and value of the Open Journal, any links they saw between the Journal and their own teaching practice (as well as their patterns of use and place of access). Students' responses indicated that many of them understood the purpose of the Journal as a means of providing insights into the thinking behind the Biology methods classes. For example:

I like to have an insight into your thoughts and purposes for the sessions; they [entries] often bring up issues that I have previously not considered or not extended greatly; it helps me think about how our ideas are different and the same; I think about how this relates to teaching and learning [and] in particular how that will influence my teaching, what is important to me; I was very interested in what you found difficult or what went differently. Value is that it allows me to see common teacher problems and concerns that you have as well. (Sally: e-mail response to Open Journal)

In a couple of instances, students explicitly linked the Open Journal with their experiences in schools.

[I] Most value your entries when I was on rounds and considering my own teaching methods. I found your diary a guide for thinking about my own experiences and abilities. I like the idea of the journal . . . I have often thought that this may be a good idea when teaching VCE classes [seniors] but I'm worried it may take too much time. (Natalie)

I think about how this [journal] relates to teaching and learning, in particular how that influences what I do on rounds . . . It's such a great thing to have insight into the teacher's aims, because when I am teaching, finding honesty from the students can be really difficult. (Linda)

Informally too, students would chat with me about Open Journal entries. On a particularly memorable occasion, a student described to me that he and his peers regularly discussed the content of the entries when they met for drinks after class!

Writing the Open Journal each week helped me to clarify my thinking in ways that may not have occurred if the journal had simply been for my private use. The expectation that the Open Journal would be published forced me to articulate my thoughts before and after a class, and to sort carefully through issues, concerns and questions related to my own or prospective teachers' learning.

In addition to learning from my own writing, students' responses to the Open Journal entries also helped me better understand their learning and my teaching practice. For instance, after one session I wrote that I had been feeling disappointed that I had not handed control of a summarising activity over to the students (Open Journal – Week 3). Soon after posting this entry, I received an e-mail from a student who described how having me summarise the session was helpful for her learning at that point in her development. She helped me to understand the session through her eyes, in a way that I had not previously considered.

> Hi Mandi,
>
> One of your comments this week was about your frustration that the last thing we left class with was your summation [of the activity] instead of allowing us to do this. I agree that as preservice teachers we would gain valuable experience by leading discussions on what we thought. However I also think that we need validation from you that we are on the right track, that what we are doing is correct. Confidence is a big factor in how we will succeed as teachers. If we are full of confidence we are more likely to be creative and daring in our approach. At this stage I am not sure who else could give us confidence other than our tutors. (Lina)

At the same time that I was prioritising my agenda, to offer insights into my thinking, (in this case publishing my disappointment about the way in which I handled a particular class situation), Lina was letting me know that her priority as a new teacher was to develop her self-confidence. She reminded me that unless she developed feelings of self-confidence then it was unlikely that she would be prepared to take risks in her own teaching. This situation illustrates how, in balancing the tension between confidence and uncertainty, the teacher educator must pay careful attention to the range of needs being addressed.

Making my thinking explicit during teaching

The other way in which I attempted to make explicit the complexities and uncertainties of teaching was through talking with my students about my thinking about teaching, during classes. My decision to make explicit my thinking was based on the idea that the problematic nature of teaching needs to be highlighted in teaching about teaching in order for careful attention to be paid to the pedagogical reasoning which underpins practice. In other words, if teaching is understood as being problematic, then in order for learners of teaching to more fully grasp the nature of the problematic, teacher educators need to articulate and explicate such situations in their own teaching. In that way, prospective teachers might begin to access, and better understand, the ideas and actions that emerge in considering teaching as problematic. This might then be a starting point for prospective teachers to begin to recognize and access the knowledge on which practice is based – but that is not always explicitly seen or understood. However, enacting my ideas was not a straightforward task. An extract from my Personal Journal illustrates my thinking about the difficulties of this approach:

> Even though I have identified that articulating my thinking about teaching during the act of teaching is an important goal of my teaching, I have also found that this is not an easy goal to 'live' as a teacher educator. I am not always consciously aware of my actions, in action, nor am I able to readily articulate my pedagogical reasoning on the spot. Usually, there is a multitude of thoughts running through my head as I teach. How do I know which of these is useful at any particular time to select to highlight for my students? (Personal Journal, Week 2)

I shared this journal entry with one of the prospective teachers, Lisa, via e-mail. She e-mailed me the following response:

Date: Thurs, 15 March 2001
From: Lisa
To: amanda.berry@education.monash.edu.au

I think it was important for us to trust that you would be able to teach us well . . . opening up your vulnerability and uncertainty about things was unsettling for many . . . It was like 'whoah! She doesn't know what she's doing all the time – holy hell! – what hope have we got?

Lisa helped me to learn that in choosing to make my thinking about my teaching available, some prospective teachers experienced a loss of confidence in my ability to successfully guide their development. Their views of my role were not compatible with the role I was enacting. Interestingly, during her initial practicum experience, Lisa deliberately chose to teach her Science classes in a way that more faithfully represented the uncertainties of its processes – to reveal its constructed, human, imperfect nature and so challenge students to think about Science ideas as other than 'black/white, right/wrong'. After I observed her teaching and talked with her following her class I wrote in my Personal Journal about the difficulties of balancing trust and vulnerability.

> Lisa brought to my attention something really important. . . . There is a distinction between helping students see that science is socially constructed knowledge and that teachers are humans who can be wrong, and trusting that your teacher knows what s/he is doing. A distinction that sounds easy but in fact can

be difficult to enact. I feel like I have perpetuated that to some extent through my being very honest and vulnerable with my students, which led L to see merit in that and do that with her students. But because they didn't really know what she was on about . . . it didn't work in the way she wanted it to. Perhaps that is the same for me? (Personal Journal, Week 12)

Another prospective teacher, Jill, told me during interview that she found my 'thinking aloud' helpful for her thinking about practice because it gave her cause to reflect on her own teaching, although she added that it was not an approach she would use with her high school students because of the difficulties associated with trusting the competence of the teacher.

> Jill: I think that the reflection within the class allows us to know what's going on and so that it's a kind of a process of "Would I do that, or would I say that?" So as an example, when you say, "Oh I've just cut that discussion off and I probably shouldn't have done that . . ." it actually gives us some insight into the way that you're thinking, and I would say "Is that something I would do in class?" . . . It's not something I would do in a VCE [senior] class because I think that students really need to know that the person out the front has some confidence in what they're doing and they make a decision for a particular reason, so I'm not sure how those students . . . would find that, but for me, it's good because I can say, now would I be thinking that at this particular time? Would I have cut that discussion off? And if I did, would I be thinking the same thing that you're thinking? Would I be thinking it was an inappropriate time to stop that discussion? (Jill Interview 1: 25–27)

Just as it was difficult for me to identify and articulate what I was thinking while I was teaching, so too these prospective teachers, immersed in the 'here-and-now' of an experience found it difficult to look beyond what immediately preoccupied them. The challenge of attending to learning on several levels is considerable and unless I encouraged prospective teachers to look around, to consider the bigger picture of their learning about teaching, then there was a real danger that they would believe that what immediately surrounded them was all there was to see (and know). I wanted to make explicit to them differences between being a learner of teaching and a 'do-er' of teaching so that they might come to better understand the ways in which these different perspectives influence practice. However, if my students saw me as somehow very different, not connected with the real world of practice, and they saw that experienced teachers in schools did not explicitly consider their teaching in this way, then what I did may have simply been interpreted as an (unnecessary) academic exercise.

If supervising teachers in schools made explicit only one role, that of 'do-er' then the power of my message was considerably reduced. This was particularly evident when one of the prospective teachers told me that there had been some discussion between members of the Biology methods class that they would like to learn more about teaching and less about how I teach. This point was further emphasized for me a little later in an e-mail from Lisa.

Date: Mon, 2 April 2001
From: Lisa
To: amanda.berry@education.monash.edu.au

Today Mandi stopped the class and explained why she was doing something and how she thought it would work. I really liked this but I wonder if there were some students who thought, "I really don't care"? . . . What I am trying to get at is: Do we need to know we are meant to be learning something to learn it?

Bill reported that my 'thinking aloud' was unhelpful for his learning about teaching because it added an extra dimension of thinking that overloaded and confused him.

Bill: We're trying to learn to teach . . . we're trying to be students and we're trying to be students of teaching all rolled into one . . . It's a funny little world, because when you are in a class sitting, watching, it's difficult . . . you're thinking of a number of things. You're thinking ahead, backwards, all this stuff; sometimes it's hard . . . it's difficult for people to think on two levels all the time. Sometimes you know, everyone's sort of going what's going on? What's this all about? I was just starting to think about the task I was doing and you're asking us to think about why you organised it that way. (Bill Interview 1: 200–204)

Teacher educators report feelings of uncertainty as they begin to enact new approaches to practice (White, 2002). These feelings can be conveyed to prospective teachers and interpreted as shortcomings on the part of the teacher educator. Deciding what aspects of practice to make explicit, how to make them explicit, and when, so that they are useful and meaningful for prospective teachers, lies at the heart of this strand of the tension between confidence and uncertainty. It is a risky business for the teacher educator and requires the establishment of a trusting relationship with the class, as I learnt. Loughran (1995) had already discovered this in his efforts of similarly 'thinking out loud' with his student teachers.

Choosing an appropriate time to explain that I would be "thinking out loud" and my purpose for doing so was important. I had to have a sense of trust in the class and they with me otherwise my behaviour could appear to be peculiar rather than purposeful. There was a danger that talking aloud about what I was or was not doing, and why, could be interpreted as lacking appropriate direction. This could be exacerbated by the fact that many beginning teachers enter the course believing they can be told how to teach. It could be a risk which might compromise my supposed "expert" position as someone responsible for teaching teachers. (p. 434)

It is interesting that Loughran's view (above) about the possibility of compromising one's position through what one selects to share with prospective teachers echoes the ideas expressed by Lisa. It seems odd, in hindsight, that I did not attend more carefully to the possible conflict between my own expectations and those of the prospective teachers in deciding to share my teaching in this way. Mostly, I assumed that my teacher educator "authority of position" (Munby & Russell, 1994) would afford me an expert status that would automatically convince my students that what I was doing was worthwhile. Alternatively, I may have assumed that what I was attempting to show them about teaching was so compelling that they could not help but be convinced by the valuable insights they were gaining into my teaching that would assist in the development of their own.

The role of trust

Trust is crucial in supporting new approaches to practice. For example, what reason would Bill have to trust that it was worth suspending his needs for developing certainty and confidence for the uncertain alternatives that I was proposing? A readiness to take a 'leap of faith' comes from trust in the teacher educator, so that even though prospective teachers may not agree with what the teacher educator is doing, there is at least preparedness from the prospective teacher to make a wholehearted effort. Andy reported in our first interview, that for him, preparedness to step out and do something different came from knowing that the teaching approaches that he had grown up with, and that had worked for him, were not necessarily helpful for his students' learning.

> Andy: Most people won't abandon what they know and what works because there's just too much risk involved, but what . . . people such as myself should do is maybe think, no don't worry about what I do know and what does work, I'll try this. I really will try . . .
>
> Mandi: What would motivate you to try?
>
> Andy: The knowledge or the information . . . at a very early stage, the awareness that what I do and what I use probably doesn't work for most people at a secondary [school] level.
> (Andy Interview 1: 61–63)

Trust is something that is developed within, and supports, a safe environment. However, in paying attention to social aspects of learning about teaching there is a need for all involved (teacher educators and prospective teachers) to trust that learning through experience is valuable and worth their efforts. This is not really possible though if the teacher educator assumes total control of the learning environment, losing sight of, or not acknowledging, individuals' needs. In my efforts to minimize my role as expert and relinquish control of the learning to the students through such activities as peer teaching, it became clear to me the importance of recognizing and acknowledging *feelings and* expectations of prospective teachers about learning to teach.

Trust (and in its wake, confidence) can be quickly undermined when teacher educators are not attentive to the emotional as well as the pedagogical needs of their students, as I soon learnt. A number of students felt very uncomfortable with my efforts to create a climate in which their teaching was open to professional scrutiny by their peers which led me to ponder how I could help prospective teachers to see learning as a collaborative venture, open to professional critique, and challenge yet not break their confidence in themselves, each other, or me. (This aspect is strongly linked to the tension between safety and challenge.)

Loughran (1997) identified three forms of trust that need to be developed for a successful teaching/learning relationship to occur. One form is self-trust, from the teacher educator him/herself. The teacher educator needs to trust that prospective teachers will view him/her as someone who can genuinely engage them in the learning process. A second form of trust comes from the learners. This involves learners feeling 'safe' to express their ideas, questions and concerns in such a way that permits challenge to these ideas but at the same time, preserves the learners' self-esteem. A third form of trust requires that issues raised within the teaching/learning context are received in a genuine manner and dealt with in a way that shows concern to resolve the dilemma. This means that the teacher educator needs to work with prospective teachers to help them address their individual needs and concerns, rather than supplying 'expert' knowledge to solve problems (Loughran, 1997).

In hindsight, I recognise that a number of the difficulties I encountered in working with Biology Methods students were related to the absence or undermining of these different forms of trust. While I was keen to develop a learning environment that engaged the prospective teachers as learners, I was not confident in my ability to do so (teacher educator self-trust). The effect of this on my teaching was that a number of the prospective teachers saw me as someone who was unsure about what she was doing. This led to a lack of trust in me, and my approach (learner trust). This made it a little more difficult to establish the joint trust that was necessary for meaningful learning to occur. And, although I worked hard to withhold my expert status as a teacher educator so that I could encourage my students to work towards developing their own knowledge of practice, for some, this resulted in a further undermining of their trust in my ability to effectively support their learning (trust that individual concerns would be dealt with in a genuine way).

SUMMARY: WHAT DID I LEARN FROM EXAMINING THIS TENSION WITHIN MY PRACTICE?

One of the tasks for teacher educators in managing the tension between confidence and uncertainty lies in being able to "nurture confidence in the worth of . . . [prospective teachers'] ideas while also encouraging them to reflect on and rethink their views" (Schulte, 2001, p. 30). I have come to learn that in teaching about teaching, being aware of the *feelings and expectations* that prospective teachers have about learning to teach is important because of the impact that the demands and expectations of the subject will have on them and their practice. Thus, confidence emerges as a search for balance as confronting some of their needs and expectations through the processes and practices of Biology methods could, if not carefully thought through, shake their confidence. This occurs in the way in which I respond to my students and as they in turn respond to the situations they encounter in the subject.

For me, coming to understand some of the difficulties associated with learning to recognize when I was pursuing my own needs and when I was responding to students' needs came out of my investigation of my practice. There are no 'scripts' to know how to behave in such circumstances. Ongoing critical self-observation of one's teaching practices requires being both observer and participant in one's teaching, a task that is not easily achieved and that requires considerable practice (Mason, 2002). Teacher educators not only need to be able to articulate a "running commentary while driving" (ibid, p. 224) about what they are doing and why, but they also need to thoughtfully select what to make explicit from their "commentary" for their students, when and why, if they are to help them confidently develop their understandings of the complex uncertainties associated with teaching.

Chapter Seven

ACTION AND INTENT

action (n) something done; a deed or act

intent (n) an aim, or purpose

> One of the hardest things teachers have to learn is that the sincerity of their
> intentions does not guarantee the purity of their practice. (Brookfield, 1995, p. 1)

INTRODUCTION

The tension between *action and intent* arises from discrepancies between goals that
teacher educators set out to achieve through their work and the ways in which these
goals can be inadvertently undermined by the teacher educator's own choice of actions
to achieve them. Named by others researching their teaching practice as, "working
within and against myself", "interactional misfires" (Macgillivray, 1997, p. 470) and,
"the tension of opposites" (Palmer, 1998), this tension focuses on better understanding
the relationship between one's actions and intentions, and attempts to bring these
toward closer alignment. This chapter explores the ways in which I learnt to recognize
this tension (and its effects) within my teaching and how recognition of discrepancies
between what I wanted to achieve and the ways I chose to act, led to the development of
my understanding of practice. (Seeking to more closely align my actions and intentions
remains however, an ongoing task.) First, I discuss the frames that inform my understanding of these ideas then, I draw upon one of these frames, that of 'deep and surface
discrepancies', to illustrate the tension between action and intent within my practice.

RECOGNIZING DISCREPANCIES

Recognizing discrepancies between my intentions for prospective teachers' learning
about teaching Biology and my actions to bring about these intentions presented a
considerable challenge. Patterns of teaching behaviour are established over a long

time (Mason, 2002) therefore teacher educators may find it difficult to see (let alone give up), habitual ways of behaving even though these ways actually work against their intended goals for their students' learning (Brookfield, 1995; Korthagen, 2001). Even after learning to 'see' my actions as a teacher educator, I found it difficult to recognize conflicts between my teaching actions and my beliefs about teaching or, that students' perceptions of my actions were different from my own perceptions of these actions. The work of Brookfield (1995) and Korthagen (2001) offers helpful frames for understanding these differences.

Assumptions and Discrepancies as a Means of Understanding Practice

Brookfield discusses the notion of 'assumptions' that are, "taken for granted beliefs about the world and our place within it that seem so obvious to us as to not need stating explicitly" (Brookfield, 1995, p. 2). He adds that, "In many ways we *are* our assumptions. Assumptions give meaning and purpose to who we are and what we do" (ibid, p. 2). Brookfield identifies three categories of assumptions: paradigmatic (basic structuring axioms that we live by); prescriptive (what we think ought to be happening in a particular situation); and, causal (if/then type statements about how different processes work). Brookfield acknowledges that becoming aware of the implicit assumptions that, "frame how we think and act is one of the most challenging intellectual puzzles we face in our lives. It is also something we instinctively resist, for fear of what we might discover" (op. cit., p. 2).

Korthagen uses the term "discrepancy" to describe inconsistencies or contradictions between action and intention. Korthagen draws on an adaptation of Bateson's (1972) 'onion model' (Korthagen, 2004) to identify different levels (or layers, using the onion analogy) at which discrepancies can occur for an individual. In this model, the outermost levels are concerned with observable aspects of teaching (e.g., context, behaviour, competencies) while the inner most levels are connected to an individual's beliefs, identity and mission (sense of purpose in the world). Korthagen asserts that discrepancies occur when these levels are in opposition (for example, a mismatch between surface behaviours and mission), and that the role of the teacher educator is to develop strong self-awareness of each level so that factors inhibiting the realization of one's intentions can be reduced. Through developing her self-awareness, the teacher educator is then better positioned to assist prospective teachers to recognize and achieve their teaching goals.

Brookfield's "levels of assumption" and Korthagen's "discrepancies" have been important influences in developing my understanding of differences between action and intent in my own practice. Drawing from these ideas and my experiences, I have used the notion of 'discrepancy' to name the mismatches that became apparent to me in my teaching of Biology methods. As this self-study progressed, different types of discrepancies were revealed. Some were relatively easy to recognise and, although surprising to me, did not shake me too hard; they were behaviours that I felt comfortable to modify so as to better fit my intentions. For example, I learnt from student feedback that the approach I took to reduce the effect of my "authority of position" (Munby & Russell, 1994) and encourage greater peer interaction, by physically moving myself to one side of the room away from the immediate focus of student attention, did not achieve the effect I intended. The students simply accommodated the change by physically shifting their positions so as to better see, and talk to, me. (This example is discussed more fully later in this chapter.) This meant that I needed to find different ways of helping to bring about the change I sought. I have named this type of discrepancy, as illustrated in the previous example, *surface* discrepancies. These were quite different from the *deep* discrepancies that I also uncovered through this self-study. Deep discrepancies were those that profoundly challenged my beliefs about myself and my approach to teaching about teaching, for example how power was distributed in my class, or uncovering 'rules' that I was unknowingly using that privileged particular students, or that put my own needs in front of those of the students. Uncovering these *deep* discrepancies presented serious challenges to my understanding of my teacher educator 'self' and did not lead to easily resolvable changes.

It is important to note that my use of the term 'discrepancy' within this chapter is not intended as a means for making negative judgments about my practice (although at times I felt very uncomfortable about some of the discrepancies that were revealed). Rather, the use of the term 'discrepancy' is intended to carry a neutral meaning; discrepancies simply exist within one's practice, and hence within this set of tensions, act as an important means of highlighting a particular aspect of my practice for study.

I now consider a series of examples from my practice to illustrate some of the surface and deep discrepancies that were brought to my attention through this self-study. First, I consider surface discrepancies by looking at particular principles of my practice, the way(s) in which I chose to enact these and how my behaviours were variously interpreted. Then, I explore some aspects of my practice that were operating at a more subconscious level; which led to deep discrepancies being uncovered.

Surface Discrepancies: Some Examples from My Practice

Example 1: Attempts to reduce my "authority of position"

As already identified, an important principle that informs my teaching about teaching is that prospective teachers should learn to pursue their own understandings of teaching and learning, not depend on, or try to please, external authorities (such as me, or their school based supervising teachers). In order to realize this principle in my practice I enact particular behaviours that I believe will enhance my students' sense of personal agency and reduce their need to depend on my "authority of position" (Munby & Russell, 1994).

One of the ways that I do this is by extending the amount of 'wait-time' that I use. In this way a range of student voices may be heard and the typical classroom scenario in which the teacher closes on a student response and quickly moves on to the next idea without creating opportunities for developing understanding, can be minimized. Another strategy that I use to enhance prospective teachers' sense of agency is by attempting to diminish the extent to which they rely on me in the classroom (for example, by limiting my teacher talk and positioning myself so as not to be the focus of attention in the room). Yet another is by withholding judgment, whereby I accept what my students say, and encourage a variety of responses without closing on correct or incorrect answers. My behaviours are the result of my purposeful deliberation to enact a belief about facilitating prospective teachers' independent development. So, it came as quite a shock to me when, early in the year, Lisa shared a journal entry with me about her perceptions of my teaching behaviours in the Biology methods class. Although initially taken aback by what I read, I began to understand some problems with my teaching approach as it was experienced 'through Lisa's eyes'. (At the same time, I was also extremely pleased that Lisa felt safe and confident enough to openly share some of her thoughts with me at this early stage.) An extract from Lisa's journal entry describes how she experienced my decision to enhance student-to-student communication in the class.

Subject: LOCATION, LOCATION, LOCATION
Date: Tues, 6 March 2001
From: Lisa
To: amanda.berry@education.monash.edu.au

Today in BIOL, we were in grouped tables and had a class discussion at the beginning. Mandi stood at the back of the class – I think so that maybe we wouldn't focus on her or maybe something to do with the [video]camera? Tables were grouped in threes, so that one was facing the front. I hurt my neck craning to look at Mandi. Although I think this was sort of the point – that she didn't want us to focus on her, we did anyway and I hurt my neck, which I am not happy about. She also didn't move around much.

Lisa's highlighting of my behaviours led me to consider the need to develop alternative behaviours that might better serve to reduce the discrepancies between the pedagogy I was enacting and my intentions. Even so, it was particularly surprising for me to learn from Lisa that my actions were not effecting the goals I had anticipated. Without Lisa's observation I would not have been aware that the problem existed, especially since I had been more focused on my goal than how it was experienced by my students.

Obtaining feedback from the class members was an important means for me to be able to see differences between what I intended my students to learn and how they interpreted my behaviours. However, few class members were as forthcoming as Lisa, so I had to find other ways of accessing individuals' perceptions of Biology methods classes. Student interviews, post-class feedback sheets and e-mail correspondence offered some ways of gathering their thoughts. While such approaches were helpful, I struggled at times to know what questions I might ask them to help me see what I could not yet see in my own teaching. Simply asking them, "How did it feel for you to be a learner in this class, today?" often met with polite, non-committal responses. I guessed that asking such questions stepped outside the normal boundaries of students' experiences of their preservice education. Therefore, they were not quite sure what I was asking them about, or why. Also, given our power relationship it was hard for them to know what was genuinely acceptable to speak about in terms of their experiences of my teaching.

One approach that I used to find out more about how I taught was to invite a trusted colleague into one of my Biology methods classes as a participant–observer. I asked my colleague to observe my teaching and to conduct a discussion afterwards with me, and the class about our various perceptions of the teaching and learning we had experienced. There were several important reasons for asking my colleague to work with the class in this way. One was that I hoped that I could gain some more insights into how I taught through discussion between an observer and the prospective teacher group (since I thought that they might be more ready to respond to someone other than me asking them questions about my teaching. Of course, the possibility existed that they might say less because of their lack of familiarity with my colleague and with this type of experience).

Another reason was aligned to a goal for my teaching about teaching; that prospective teachers be presented with opportunities to see that teaching consists of more than what is immediately apparent on the surface. Through the process of being asked questions about my teaching I anticipated that these prospective teachers might learn something about the 'why' and 'how' of my decisions for their learning. And, by explicitly 'unpacking' the process of pedagogical decision-making with them, I hoped that at least some of them might have been stimulated to think about the nature and complexity of the decisions they made in their teaching, as well as how particular decisions they made in their teaching impacted their students' learning. My intentions for the session then, were framed around ways of helping me, and my students, 'see into' practice. In the following example, the events of the session, together with the various perceptions of, and responses to what I did, are described.

Example 2: "Why would we want to learn about Mandi's teaching?"

Immediately prior to the Biology methods session that my colleague visited I wrote in my journal:

> . . . I must admit to being rather apprehensive about the whole thing. John [colleague] doesn't know what I'm teaching and he doesn't know the group. I do not know what he is going to ask me [about my teaching] though I've tried to predict a few questions in advance. Yet I do believe that this is a worthwhile risk to take since I think it can help us to have new conversations within the group about teaching and learning in biology. (Personal Journal, Week 4)

After the session, I asked students to comment on what they saw as the purpose for conducting the session the way that I did. Lisa's response focused on the idea of 'practising what I preach' as a teacher educator: *"I really like the fact that you are doing something you want us to do . . . Other teacher [educators] say, "You do this or you do that" and they don't do it themselves. You're taking a risk which probably isn't very comfortable and you're wanting us to do it so you're doing it first"* (Video transcript). Another member of the class, Ellie, felt she had been shown that *"even experienced teachers struggle"* with their teaching, and the session gave her an idea of the way in which she might discuss her teaching with her practicum supervisor during her next field experience. Other class members interpreted my intentions differently. One student, Joanne, expressed her difficulty in trying to attend to thinking about teaching while still puzzling over the Biology content ideas presented in the session (*"I'm still thinking about antibodies and you're talking about teaching"*). Joanne's comment echoes Bill's experience (discussed in the previous chapter), about the considerable and complex demands associated with attending to learning about teaching on several levels concurrently (i.e., Biology content, experiences as a learner, teacher behaviours).

Several of the prospective teachers left the Biology methods session that day saying that they would like to, *"learn more about teaching and less about how you (Mandi) teach"*. (This information was conveyed to me through informal conversation with a couple of the class members.) Given the timing of this session (a week before the first practicum) and the concerns of many of these prospective teachers to learn all that they could to be successful in the classroom, this latter response was not surprising, albeit disappointing for me to hear. Their experience helped to remind me that people's actions are interpreted according to the 'frame' of their experience. In this case, the needs and concerns of these prospective teachers about their impending teaching experience, i.e., what and how to teach particular Biology topics to senior high school students, framed their learning from the session. Despite the fact that within the session they had experienced several teaching activities that they might have used with their students, this seemed to have figured less in their memory of the experience than the discussion of my teaching.

It was interesting to me then, that later, during the second round of cohort interviews, Sue made an unprompted link between the impact of my colleague's visit on her thinking about teaching and the difficult but necessary task of identifying why she used particular teaching approaches in her classes on practicum.

Sue: You [Mandi] normally explain why you're doing something and I find that really helpful. Like that day that John [colleague] came in and asked questions. And he asked what you were doing when you were walking around [the room] and you really had to explain what it was. I think it was checking [task progression]? I knew that, but I thought, yeah, to an outsider you would just look like you are taking a back seat. I found that really interesting that he was there that day. And from that point, you've continued to explain why you're doing things.

Mandi: And that's helpful for you?

Sue: Yeah . . . You know, there's a question on the lesson plan [prestructured format], why am I doing this? I hated that question but I really thought that if you can't answer it you shouldn't be doing it, so I really thought about why am I doing it? What is the whole point in doing it?
(Sue Interview 1: 240–244)

Sue's words were helpful for my understanding about the effects of opening up my teaching for discussion. The Biology methods session that day impacted students' thinking in more ways than I had initially recognised. Sue's response highlighted two important aspects of my experiences of the tension between action and intent. First, my impression of students' perspectives of that particular experience was gathered from a selection of individuals. I was not privy to the thinking of each student, nor all of the thoughts of any one student. Therefore 'seeing myself through these students' eyes' was inevitably an incomplete view of my actions. (This is not to say that these views were not useful, rather it was being reminded of this incompleteness that was significant for judging how my actions were perceived.) Second, and more importantly, I needed to remember that prospective teachers' interpretations of my actions and the ideas stemming from them about teaching and learning, may change over time as their needs and concerns, and hence their frames for thinking about teaching, also change.

The tension between action and intent is played out over time as experiences are had, recalled and reinterpreted by all participants. Learning from experience may not be apparent to the teacher or student at the time; it may not come to fruition until later. Therefore 'sowing seeds' for learning and trusting possibilities for future learning is an important teacher educator competency that impacts this tension.

Example 3: Different interpretations of a 'straightforward' task

A third example of a 'surface discrepancy' emerged in the same Biology methods session (colleague observation). This time, the discrepancy arose from a surprising difference between what I intended students to experience from a particular activity and what a number of them actually experienced from the activity. Prior to the session, the class had discussed approaches to eliciting individuals' prior knowledge of a Biology concept in order to plan purposeful learning experiences that built upon this prior knowledge. This led me to introduce a diagnostic assessment tool that prospective teachers might consider using with their students for this purpose. The content focus of the session was disease transmission. At the beginning of the session I distributed to each student a list of terms commonly associated with learning about disease transmission, for example, infectious disease, parasite, pathogen, vector. Students were instructed individually to write about the meaning of each term, as they understood it, so as to find out what they already knew, and to compare this prior knowledge with what they learnt during the lesson. This diagnostic activity was, as I saw it, a straightforward and non-threatening task. And, I had no evidence to believe that when I observed individual students doing the task that they felt other than comfortable. In our discussion following the teaching, my colleague asked the class about their experiences of this tool. It came as a considerable surprise to me to hear a number of them confess that they had found the initial part of the task threatening!

This example again highlights how our frames of reference for interpreting experience influenced learning expectations and outcomes. The following exchange, an excerpt from my in-class discussion with my colleague about the task, illustrates this idea:

Mandi: What I was trying to do was create a private way that people could write down their ideas so they could come back to it later. Because in the last couple of weeks we've been talking about how do you keep track of an individual's progress? And I thought, here's a way that I could introduce a procedure, something that you might use as a way to keep individual track [of students], but now I can see that it can have a reverse effect . . .

John: . . . people interpret it differently . . . Because I experienced it, it became an issue. In watching what happens in teaching, it's different to watch to be a learner in a situation. I was really pleased with what you said R [referring to student comment about the tool]. We heard the same thing differently. The words were the same, the intent was the same, we reacted differently to it.
          (Video: Week 4)

This incident highlights differences between my teacher educator perspective of *introducing* the diagnostic tool and prospective teachers' perspectives of *experiencing* the tool and how these perspectives can radically differ because of the frames each person brings to the experience. My teacher–educator frame acknowledged the sound pedagogical purpose of the tool. I anticipated few, if any problems since prospective teachers were 'simply' required to do it – i.e., privately write about what they knew of the Biology concepts. Prospective teachers' perceptions of the task as 'do-ers' included, for at least some of them, feelings of vulnerability as learners whose content knowledge was placed under scrutiny, even if in a private way. Because I had not been in the position of experiencing what it actually felt like to do the task, and there was no risk involved for me, I didn't imagine that I was creating a potentially uncomfortable situation for them. My awareness was limited by not taking another's (i.e., the students') perspective. My colleague, and a number of the prospective teachers, reminded me that this discrepancy occurred in my practice because my actions, although pedagogically aligned with my intentions, did not acknowledge the possible alternative learner perspectives of the task. While the task was a perfectly reasonable activity to include in the session, I needed to be mindful of my students' needs so that they might be discussed, explored and compared with the ways in which high school students might be likely to experience the same task.

Sometimes actions do not coincide with their intentions whilst at other times an action and its intent can be more closely aligned. In a final, brief example, I illustrate how I came to recognize that my intentions for students' learning about wait-time had been realized, at least to some extent, through my actions.

Example 4: When action and intention coincide

In an earlier chapter (Telling and Growth), the example of wait time was identified as something that Nick had learnt from the Biology methods classes. While Nick may have been simply trying to please me in talking about the use of wait time in his own teaching, it is clear that he understood the message about wait time that I had modeled for the class. Interestingly, he also remembered that I had explicitly discussed this aspect of my teaching, including my pedagogical reasoning in class. Through making explicit my reasoning to students in this way, including the complexities and decisions associated with my practice, I hoped that my students might be prompted to examine more closely and learn about their own pedagogical decision-making. While there is little evidence to suggest from the above example that Nick had developed the capacity to do this, his awareness of the importance of wait time was raised through my actions.

Uncovering surface discrepancies in my practice provided an opportunity for me to modify my actions in order to better achieve the purposes I intended. Uncovering deep discrepancies however, presented a much more challenging task because of the more subconscious levels at which the discrepancies occurred and the corresponding degree of self-awareness required for me to recognize and accept them. I now turn to an examination of some of these 'deep discrepancies' uncovered through the self-study of my Biology teaching practice.

Deep Discrepancies: Some Examples from My Practice

Example 1: An "interactional misfire"

An example of a situation that significantly challenged my view of my actions compared with others' perceptions of my actions, occurred during Bill and Joanne's peer teaching session. Bill and Joanne had chosen to teach the class about cell structure. Following some initial content input, they organized the class into small groups then asked each group to construct a cell membrane from materials provided. I joined one group and as our group began work, I quickly recognized a commonly occurring classroom situation whereby the more capable and confident group members immediately took control of the task and those less confident members quietly sat back and watched. I thought this was worth paying attention to since prospective teachers would also encounter this situation in their teaching, so I asked the group if we could modify our approach so that those of us (me included) who felt less comfortable with the content knowledge required for the task could be the do-ers, and those who already knew the relevant content could support us (without simply telling us the answers).

My intent was to model an approach that these prospective teachers might consider using with their students in a similar situation, and to highlight to Bill and Joanne an important issue for teachers to consider in setting up group tasks. Because of the way I reconstructed the task, our group spent quite some time talking about the membrane structure and did not complete all of our assigned work. During his regular visits to our group Bill expressed his frustration that we were not making progress with the task in the way he had anticipated. I tried to reassure him that we were doing our best, but, as I later learnt, he clearly believed that I had another agenda, that was to sabotage his teaching efforts. Viewed through the lens of this tension between action and intent, the situation is a good example of "an interactional misfire" (Macgillivray, 1997, p. 479). In this case, my own need to illustrate a teaching problem overtook my understanding of Bill's need to have a successful teaching experience. Consequently, my attempts to influence Bill's learning 'misfired' compared with my intentions. An extract from our discussion of this episode (during my second interview with Bill) is included below:

> Mandi: It wasn't that we were neglecting the task. In fact we were thinking about it really hard, talking about how the chemicals were joined and what they were . . . and so we didn't even get time to get to the [work]sheet because we were too busy working out what we didn't and did know. It was a really good task for us and we could have spent half an hour on it just talking about it . . .

> Bill: Yeah, but the thing is you didn't have half an hour. The time limit was written on the sheet and I mean that is part of doing an activity; knowing what you're supposed to do, knowing what your materials are and looking at how long you've got to do it. So if after 9 minutes, you haven't done anything, you haven't done the activity, the idea is to get something together . . .

> Mandi: Ok so here's a really good thing, I think, that I want you to think about . . . Is it the process or the product that you're after? Our process was A+; our product was an F.

> Bill: But your product was going to be our proof of teaching.

> Mandi: Aah, now if . . . the 2 smart people out of the group of 5 had built the model, your proof would be that you'd done a really good job. But could they have built the model before you even started doing your teaching? And, how would you know?

> Bill: (pause) Well, good point. I don't think they could . . . I don't know that people would have known the intricacy of that phospholipid bilayer before.
> (Bill Interview 2: 360–394)

Bills's comment that the group product was intended to serve as a measure for his and Joanne's competence in teaching about cell structure ("*your product was going to be our proof of teaching*") tells me that it is little wonder he interpreted my actions as 'sabotage' rather than instructive. In this case, the discrepancy occurred at least to some extent, because the approach I used to enact my pedagogical intentions, although "sincerely rooted in a pedagogical idea" (Korthagen & Vasalos, 2005, p. 54), did not invite Bill's participation since his primary concern was in establishing his identity as a successful teacher. I was unaware of how these different factors were interacting, which therefore considerably limited the effect of my approach. As a consequence, a form of power struggle ensued between us as each attempted to assert the validity of our actions.

Example 2: Uncovering my 'tacit rules' for student behaviour

A second example of a 'deep discrepancy' was revealed when Lisa helped me to uncover some of my 'tacit rules' governing the ways that I wanted prospective teachers to interact with me when they sought assistance. Although I was aware that I acted in particular ways when students sought support from me, I hadn't realized the extent to which my behaviours privileged particular kinds of learners. It wasn't until I carefully considered comments from an e-mail sent by Lisa that I came to understand that I was unconsciously applying a set of rules in these interactions. Recognising this aspect of my approach led to new and surprising insights for me.

From Lisa's feedback I learnt for instance, that if students wanted support from me, they needed to approach me personally, usually outside of teaching times, they needed to demonstrate that they had already gone to some effort to find out what they wished to know (usually this was ideas about how to teach particular Biology topics) and they needed to appear very thoughtful and reflective about their teaching. (Interestingly, this set of 'rules' is very similar to what both Macgillivray (1997) and Tidwell (2002) found in their studies of their interactions with their students.) Because of my strong belief that prospective teachers develop their own ways of teaching rather than learning to mimic mine, I deliberately kept to a minimum the supply of 'engaging activities' that I provided in Method sessions; although there were some students I did supply such resources to, one of whom was Lisa. Prospective teachers who received these resources were most often the very students who least needed my support to develop engaging lessons. Typically these prospective teachers had come to talk with me privately about their teaching, to reflect on issues that were concerning them, and in the course of our conversation I would spontaneously offer them teaching resources that they might find useful and engaging for their students, often without them even having asked! Those student teachers who didn't know my 'rules', didn't get the same access to resources, even though they may well have benefited most from this kind of support. In effect, I was maintaining a public principle (to withhold resources so that students might be prompted to develop their own) that operated differently in private, and with certain types of prospective teachers.

This situation was brought to my attention via an e-mail conversation with Lisa. She asked why I had given her a "cool idea" to support her teaching of genetics but I hadn't offered it to any of the other students. She was sure they would have found it enjoyable and useful too.

Date: Mon, 22 August 2001
From: Lisa
To: amanda.berry@education.monash.edu.au

I have another question – Why didn't you give the Vegie People [genetics activity] out to others? Will you give it after this [teaching] round? I can see why to some degree, but kind of find it hard to see why if you have all these cool ideas, why you don't share them with us. Do you think/hope we will be exposed to some of them anyway? . . . And how come I get it? Do others get things? Is it only the people who come to you that get rewards? What about the solitary workers and shy people? Still, I guess that's life isn't it – if you don't ask, you don't get, kind of thing. Hope this doesn't sound critical because it's not – just am fascinated a bit by this. But how do you teach us to ask if that's what you want to do? I think I am not making sense here – sorry.

Lisa's question, *"But how do you teach us to ask if that's what you want us to do?"* directly targets a problem for me in my teaching about teaching. It wasn't my explicit intention to teach my students how to ask me for help – my public view at least, was that students should know how to help themselves. What this situation suggests though, is that I was rewarding particular ways of asking for help and so privately (and tacitly) teaching these students about differences between what I said and what I did. Moreover, the kinds of students who worked out my 'rules' seemed to be those who were most able to find resources for themselves anyway (i.e., not 'solitary' or 'shy'). In fact, in hindsight, those who successfully sought my help were most like me, keen to succeed and open to asking for help in achieving their success.

Example 3: Intent and relinquishing control

A different way in which my rules for student behaviour became obvious and, in the process, uncovered another deep discrepancy, occurred during a Biology methods class in which I asked Ellie, one of the prospective teachers, to run a debrief discussion of a peer teaching session. Ellie and I had previously shared some private conversations about her learning about teaching and I was pleased with the deep levels of reflectivity that she had shown about her own and others' teaching. I thought that by giving her a chance to run a debriefing discussion this would challenge and extend her capabilities as a teacher and that, at the same time, her thoughtfulness would provide a helpful model for other students. However, what I did not recognize until later, was that my expectations were for Ellie to 'be like me' – to run the debrief as I would have done – and I was disappointed when she was unable to do so in the manner that I expected.

I illustrate this incident in the form of a 'vignette' that draws on the rich variety of data (videotape of this lesson, my recollections, my journal reflections and an audiotaped conversation between my colleague and myself a few days after the incident) that was available to me about this incident. Hopefully, in so doing, a more meaningful sense of the various perspectives of those involved can be conveyed.

Vignette:	*Handing over and taking control*
Mandi:	Ellie, can you please lead the discussion?
Ellie:	What? Can I lead the discussion?
Megan:	Just say, would anyone like to make a comment?
Ellie:	Okay. What did everyone think that they [teachers] were trying to teach us about today?
Bob:	Lollies!
Ellie:	Are there any comments about it?
Tash:	I thought it started off well . . .
Simone:	I thought it was good how the teachers didn't just tell us the answers . . .

As I watched the first few minutes of the discussion my frustration rose. Over the last two weeks I had worked really hard with this group to get them to talk more thoughtfully about the relationship between teaching and learning; not just to say things like "it was good". This debrief was supposed to be an opportunity for Ellie to develop her skills at running a discussion but she wasn't getting much to work with. She would have to push hard to move the comments past the level of the superficial. All of a sudden, and before I realised it, 'my mouth was open, and I had interjected with a question to the group.

"Simone, why was that helpful for your learning?"

Damn! This was meant to be a chance for Ellie to lead the discussion and for me to shut up. No sooner had I uttered these words than I was instantly aware of several people casting me looks that clearly said, "Back off. This is Ellie's discussion". I immediately did so, at the same time miming "sorry" to Ellie.

A different student then picked up the conversation.

"I think Ray and Joelle did a really good job. I didn't lose interest the whole time and I think it wasn't information overload at all and I honestly wouldn't have changed it if I was to present it to the class. It was really well paced and I liked the CDRom you used. Where did you get it from?"

I looked at Ellie and I knew that she saw what I saw. It was going to be tough to move this discussion away from student teachers affirming each other. Not that that was such a bad thing in itself, but to really start to challenge and extend their ideas about teaching and learning, they needed to explore different, more thoughtful, avenues. Again, I sat back and watched as Ellie strived to encourage more of her peers to talk about some of the difficulties and issues that had been raised for them from the peer teaching episode they had just experienced. Ellie persisted for several minutes, until it appeared to me that the debrief was 'dying' and that I should resume control of the discussion. Once more, I opened my mouth to speak . . .

Outside in the corridor after the session, an upset and angry Ellie confronted me with her feelings about my behaviour in taking over her discussion.

"Why did you do that?"

"What?" I asked, genuinely surprised.

"Why did you give me the chance to run the discussion and then take it away from me right in front of everyone?"

"I thought you had had enough. I thought I was helping you because the class didn't seem to be giving you any more ideas to work with. I'm so sorry."

"I was fine. There was no problem. You shouldn't have done that to me." She walked away tearful while the implications of my decision to 'help out' became vividly apparent to me.

Later that day, while discussing this incident with a colleague, I gradually came to an awareness that what had occurred in that class was less about me supporting Ellie running a discussion, or helping students to recognise particular difficulties with their choices of teaching approach, than it was about me satisfying my need that my students should see things as I see them and do as I would have done.

From my reflection on this incident I came to recognize one of my deeply held assumptions: when I have done something with students (in this case, modeled a debriefing discussion) I expected them to know how to do it themselves, straight away. This implicit assumption contrasts with my expressed belief about learning as a cumulative process. Although I believed I thought about learning as cumulative, I was actually searching for it to be immediate and readily apparent in the actions of my students. This discrepancy was the source of my frustration and hence, resulted in my decision to intervene in the debrief.

Also, in a similar way to the 'sabotage' episode with Bill (described previously), I neglected to acknowledge Ellie's need to experience success as a discussion leader and what success might look like for her. As part of her developing identity as a teacher, Ellie wanted to feel capable in this new role yet my need for her to act competently in a particular way (my way) considerably reduced her possibilities for success. Again, my reduced self-awareness limited my ability to perceive others' interpretations of this episode, which then limited the effect of my actions. The discrepancy emerged because on the one hand, I wanted to encourage prospective teachers to develop and grow as thoughtful Biology teachers and on the other (at a deeper level) I was undermining their efforts by setting up unrealistic expectations of what they were able to do or, holding them back if their ideas about teaching did not conform to my own. Clarifying these as problems then revealed the inner tension that was operating for me (Korthagen & Vasalos, 2005).

SUMMARY: WHAT DID I LEARN FROM EXAMINING THIS TENSION WITHIN MY PRACTICE?

Examining my practice through the lens of action and intent has given me a chance to recognize new aspects of my behaviours as a teacher educator and insights into how others perceive my actions. I understand that my intentions for prospective teachers' learning led me to enact particular behaviours that were, at times, at odds with my intentions. Sometimes what seemed to me to be appropriate ways of achieving a goal ended up jeopardizing my goal. Sometimes, my lack of awareness about the ways I was behaving left me unaware that I was unwittingly sabotaging my own goals. This was highlighted for me particularly through practices I employed that reinforced assumptions about how students should behave with me.

When a student behaved in ways that were contrary to my expectations, I became immediately drawn into this tension. This often came about as the result of my impatience for immediate and observable change in students and in myself. I am reminded through these experiences that the processes of learning take time and while my goals for prospective teachers' learning may be realized, what is most important is to remember that over time, different aspects of growth may be apparent as new opportunities for learning are presented.

In closing this chapter, I draw from Brookfield (1995) who identifies the importance of the teacher/educator having a clear sense of what s/he wants to achieve so that s/he can work from a basis of 'informed actions'. Having a sense of the relationship between one's assumptions and the practices one employs is therefore, vital. Brookfield (1995) underscores the importance of being able to see into experience from more than simply one's own perspective.

> An informed action is one that has a good chance of achieving the consequences intended. It is an action that is taken against a backdrop of inquiry into how people perceive what we say and do. When we behave in certain ways we expect our students and colleagues to see in our behaviours a certain set of meanings. Frequently, however, our words and actions are given meanings that are very different from, and sometimes antithetical to, those we intended. When we have seen our practice through others' eyes, we're in a much better position to speak and behave in ways that ensure a consistency of understanding between us, our students, and our colleagues. This consistency increases the likelihood that our actions have the effects we want. (p. 22)

Chapter Eight

SAFETY AND CHALLENGE

Safety (n) freedom from danger or risks, affording security.

Challenge (n) a demanding or difficult task; a summons to take part in a contest or a trial of strength.

> Agreeing to let people only learn in a way that feels comfortable and familiar can seriously restrict their opportunities for development. (Brookfield, 1995, p. 59)

INTRODUCTION

If, as Brookfield claims, people's learning is "seriously restricted" by their working in ways that are "comfortable and familiar", then it would seem to follow that experiences that are *un*comfortable and *un*familiar should enhance opportunities for the development of learning. The tension explored in this chapter emerged from my efforts to enact a pedagogy that was intended to shift both prospective Biology teachers and me away from the safety of the familiar and towards new possibilities for our professional growth. The tension embedded in this experience lies in engaging prospective teachers in forms of pedagogy intended to challenge and confront, and pushing prospective teachers too far beyond their comfort zone for productive learning to occur.

The title of this tension is drawn from the work of Korthagen et al., (2001, p. 75), who identify the importance of maintaining "the balance between safety and challenge" in learning to teach. In this chapter I consider how the tension between *safety and challenge* played out in my practice and in prospective teachers' learning about teaching Biology. I begin by considering the nature of teacher education as a conservative enterprise, including the interactions between prospective teachers' expectations of learning to teach and the practices they regularly encounter in their preservice education. Then, I describe the ways in which I sought to challenge expected notions of learning to teach within the Biology methods class and how I came to recognize the possibilities and limitations for learning that are embedded in an approach that operates from confronting and challenging others. Finally, I examine the effects of my approach in the interactions and learning that took place in the Biology methods class for all participants. In illustrating the tension between safety and challenge within this chapter I focus particularly on the activities associated with the peer teaching experience. This is because it was through the experience of peer teaching that the tension was most strongly 'felt' by all participants and consequently, where my learning about teaching about teaching was most vivid and significant.

Challenging the 'Safe' Practices of Teacher Education

It is a well-known idea that the teacher education practices encountered by many prospective teachers tend to support their expectations of teaching, including how teaching is conducted, the roles of learners, etc. Such teacher education practices serve to reproduce the "known" and reinforce "a culture of consensus" about teaching (Segal, 2002, p. 161). Because prospective teachers' expectations of learning to teach are rarely challenged in their teacher education, their learning about teaching, at least at university, is a relatively safe and comfortable experience (Britzman, 1991). Also contributing to their sense of safety and comfort is the way in which the normal rules of adult behaviour usually apply in this setting. Interactions between adults (particularly in western cultures) generally conform to an unspoken 'code of politeness' that permits honesty only in the expression of positive emotions and encourages courteous, compliant behaviour (Russo & Beyerbach, 2001; Warren Little, Gearhart, Curry & Kafka, 2003). This code then, guides the ways that student teachers and/or teacher educators speak about each other's practice so that "push[ing] the edges of boundaries" (Russo & Beyerbach, 2001, p. 75) such as challenging one another's views or opening up one's practice to the scrutiny of others, is generally avoided.

Approaches to teaching about teaching that encourage prospective teachers to question the underlying assumptions of the processes of learning or engage in honest discussions about the impact of teaching on the development of learning, confront these usual rules about maintaining the status quo and are therefore unlikely to be a comfortable experience for prospective teachers, or teacher educators to engage in (Berry & Loughran, 2002). Deciding to teach in ways that challenge and confront 'normal practices' not only positions teacher educators in new and uncertain roles but, also disturbs existing power relationships with prospective teachers. Segal (2002) drawing from Ginsberg (1988) warns of the possible consequences for teachers and students engaging in such new practices: "Students will be placed in the position of publicly questioning the practices of instructors who may hold the keys for their projected careers as well as discussing their own actions and statements and those of their peers" (Segal, 2002, p. 160).

My ideas about learning to teach strongly aligned with the view that avoiding uncomfortable situations minimised possibilities for learning. Consequently, in the Biology methods class I chose to work in ways that did not support traditional norms of 'polite compliance' but instead sought to provoke and disturb prospective teachers' thinking, so as to encourage them to try out new and unfamiliar ideas and practices. I anticipated that in so doing, these prospective teachers might begin to develop the confidence to imagine and enact approaches to their teaching that moved beyond expected actions and routines, and that genuinely explored their own and their students' understandings of Biology.

What Led Me to See Value in Discomfort?

My belief in the value of an approach that disturbed rather than affirmed my students' expectations of learning to teach was profoundly influenced by my experiences of developing and teaching a third year Bachelor of Education subject in the teacher education program at Monash University. As a consequence of my experiences of this subject, I decided to incorporate aspects of the approach used in this subject, into Biology methods. In particular, one of the central activities of this subject involved students planning and carrying out extended peer teaching experiences in small groups. Each group was responsible for collaborating in the planning, teaching and debriefing of a forty-five minute peer teaching session.

The peer teaching was structured in such a way so as to create an environment that supported professional critique from peers and lecturers about the teaching, which then became an important factor informing prospective teachers' development of their own teaching. The environment was carefully scaffolded to provide an experience whereby the teacher educators modeled the debriefing process first, through engaging in a critique of their own teaching, then gradually built up to the students' critiquing each other through a series of small group activities based on giving and receiving feedback about teaching. The response from students about this unit was unanimously strong – they found it useful, worthwhile and challenging for their learning about teaching (Berry & Loughran, 2002). Hence the combination of positive student feedback and my own sense of the pedagogical worthwhileness of the approach to learning about teaching within the subject, led me to feel confident that these experiences and this approach would effectively transfer into other areas of my teaching, in particular Biology methods.

In the design, implementation and subsequent learning from the third year B.Ed. subject, we (teacher educators involved) recognized the importance of maintaining a balance between *safety and challenge* for ourselves and for the students with whom we were working. We wanted to help students to be critically aware of significant features of their experiences so that they could better understand their perceptions of given teaching and learning situations. We recognized that it was not just their self-esteem at stake, so too was our credibility as teacher educators. Therefore, students needed to know that we genuinely cared about them. At the same time, we wanted them to feel uncomfortable enough about their practice to begin to examine the implications of their teaching decisions and actions. Clearly, possibilities for being hurt and making mistakes were real for all of those involved (see Berry & Loughran (2002) for a detailed discussion of this work). These elements of care, credibility and challenge were essential aspects of our approach to this subject. The source of the tension described in this chapter emerged from my experiences of re-learning and re-negotiating the balance between these elements within the context of a different subject, i.e., Biology methods, as I attempted to engage prospective Biology teachers in new approaches to learning about pedagogy.

Peer Teaching as an Occasion for Facilitating Learning in Biology Methods

During the peer teaching in Biology methods I sought to facilitate an environment in which prospective teachers could raise for themselves, and others, aspects of their experiences as teachers and learners in the situations they created. In other words, I wanted to bring to life what Shön (1983, p. 42) called "a reflective conversation" with the situation, an "on the spot surfacing, criticising, restructuring, and testing of intuitive understandings of experienced phenomena". Such an approach was intended to help prospective teachers to become more powerfully aware of their own behaviours as teachers, and the effect of their behaviours and choices for learning on the learners. The situation I set up for the Biology methods students followed a format whereby pairs of students taught the class for 45 minutes. This was followed by a 15 minute debrief in which the learning about the Biology content, the approach to teaching and the students' responses to it were discussed amongst the class. The main difference between the peer teach in the B.Ed. class and Biology methods was that Biology students were required to choose an area of Biology content to teach, whereas in the third year class, students were free to choose any content with which they felt comfortable.

Unfortunately, the peer teaching experience emerged as one of the most difficult and controversial activities of Biology methods. A number of students found the experience of teaching their peers and having their teaching debriefed, threatening and unproductive. Instead of the outcome that I had intended, whereby students might be prompted to re-evaluate their ideas about the teaching/learning process and/or their teaching approach, it seemed that for at least some of the prospective teachers, the peer teaching experience led them to maintain and perhaps even reinforce, the models of teaching they already held (i.e., teaching as telling). At least two factors contributed to this outcome: (i) my eagerness to implement this 'confrontational' approach, which led me to overlook important needs of the students to help them feel safe and ready to engage in the challenges of the process; and, (ii) the dynamic of the Biology methods class in which several dominant students appeared to have a negative impact on the rest of the group and hence reduced feelings of safety for some members of the class.

Group Dynamics and Feelings of Safety in Biology Methods

Several of the prospective teachers from the cohort that I followed during the year commented that Biology methods had a number of dominant personalities that impacted negatively on group dynamics. Kelly talked about *"so many dominant characters* [in Biology methods] *that sets up a particular dynamic"* while Andy referred to these students as *"strong personalities"* and suggested that perhaps such a personality was a function of, *"already having that teacher's kind of mind set"*. (I presume this meant one that was quite controlling.) When I asked Andy about the influence of such types on his or others opportunities to participate in class he replied, *". . . if someone doesn't feel 100 % comfortable with it, they should toughen up. (He laughs.)"* While Andy claimed that he was not unduly influenced by the dominating behaviour of his peers, there were other students who were annoyed or frustrated by it. For example, both Sue and Jacqui found it frustrating that particular class members expected their voices to be heard, yet at the same time, were unwilling to listen to others (Sue), or because of the *"nit-picking"* approach to questioning others that some took (Jacqui). The negative class dynamic was also a concern that I shared. While some of the actions I took to address the problem and minimize its impact were helpful (for example, at times restricting contributions to one per student so as to reduce the air time of dominant individuals, or providing examples of positive rather than critical language), some feelings of unease persisted between some members of the group.

Not all Biology methods students experienced the class dynamic in the same way. For instance, during our first interview, Ellie said that Biology methods classes seemed (at least initially), a "safe" place where she could not only express her ideas and opinions, but also where she could take risks and make mistakes without fear of being labeled "wrong". This was a new experience for her, one that she enjoyed and appreciated.

> Ellie: . . . it's only been during this subject that I've actually put up my hand and given my opinion . . . I've never felt safe to do that sort of stuff in a classroom, like you'd be told you are wrong or that's a wrong opinion to have. But you feel sort of safe in an environment where you can just chuck things out there . . . It's sort of a safe place to make mistakes. (Ellie Interview 1: 60–61)

When I asked Ellie to identify elements of the class that helped her to feel safe, she did not answer my question directly but, instead, mentioned the difficulty of dealing with peers who expressed a different point of view and who continued to hold steadfastly to their ideas during discussion.

> Mandi: Can you think of anything that might have helped to make it safe?

> Ellie: . . . I find that it's hard when you're having a discussion and someone's got the totally opposite opinion to you and they're not willing to concede anything . . . [but] . . . I think it feels pretty safe at the moment. (Ellie Interview 1: 62–66)

Taken together, these prospective teachers' responses offer a picture of Biology methods as a potentially risky learning environment. Concerns about the ways in which one's ideas might be judged or responded to by peers created a sense of uncertainty for some. Their responses helped me to understand why the peer teaching experience presented a daunting task for at least several of them, in terms of the teaching and the debriefing. In the peer teaching, prospective teachers were being asked to step outside their comfort zone, to teach and then openly discuss, their experiences as teachers and learners. To consider doing so, they needed to feel the trust and caring support of their colleagues. Even though I had previously identified caring as an essential element of this approach to learning about teaching (in particular, from my experiences of the third year B.Ed. subject), I had considered this aspect mainly from my own point of view. In other words, I needed to show caring towards the students. In my eagerness for these prospective teachers to have particular experiences in Biology methods I overlooked the importance of the prospective teachers' contribution towards the creation of a caring environment. My role lay in helping them create such an environment (through nurturing trust in their peers), so as to enable the prospective teachers to be ready, and interested, to engage in these challenging new experiences with me.

"Probably One of the Worst Classes . . .": Discomfort and Learning in Peer Teaching

A vivid example of how the demands of the learning environment influenced possibilities for prospective teachers' learning occurred in the first peer teaching session. Robert and Jake chose to teach the group about genetic inheritance. From the beginning of their teaching session Robert and Jake experienced challenges to their authority as teachers via some colleagues' persistent questioning of their approach. The ways in which Robert and Jake chose to deal with these challenges and the subsequent effect of this experience on different participants illustrate the difficulties associated with attempting to create opportunities for new learning and the sense of personal vulnerability that accompanies learning to teach.

What follows is an account of Robert and Jake's peer teaching as seen through the eyes of one of the prospective teachers, Kellie and myself. This account begins with an extract from my initial interview with Kellie, as we discuss the events of the session and her response to them. I follow Kellie's account with my own response to the episode, and then consider the learning about teaching that emerged for both of us in the light of the tension between safety and challenge.

Kellie: . . . probably one of the worst classes . . . was yesterday. I just felt so sorry for Robert and Jake that I just wanted everyone to leave them alone. . . . I didn't think they should have started with "I'm Mr. . . . and he's Mr. . . ." "I think they should have been a bit more relaxed and said, "Okay this is what we're going to do today". Sort of more relaxed, not like we were in a [school] classroom. I felt uncomfortable straight away and when Jeff [student] asked the difference between 'genes' and 'jeans' . . . That just really pissed me off. And then I felt that people were being quite rude at some points. Like arguing . . . They [teachers] were trying really hard and like you said, they put themselves under so much pressure . . . running a discussion is very difficult. Maybe they could have tried that at the end but not right at the start, but I felt really uncomfortable the whole way through it and I just thought, I don't want to do mine [peer teach] next week.

Mandi: Did you ever feel that you wanted to say something [to the class]?

Kellie: Yes, a lot of the time and at one point I said to Lauren, "Can you stop being so rude?" Like, in joking terms . . . At one point I wanted to tell everyone to shut up . . . I felt Jake was getting really defensive. He was getting really aggro[1]. Robert was quite relaxed, but Jake was feeling he was under attack, which he was a little bit . . . I can't imagine myself ever wanting to be up there and having that kind of response.

Mandi: How does that influence what you are going to do [in peer teach]? . . . You said, "I don't want to do it . . ."

Kellie: I certainly would never want to get up and do a discussion like they did, not in that class . . . I did notice when you [Mandi] were doing the discussion [afterward] that it was the only time there was a controlled discussion. And you actually had to say to someone at one point, "No, I don't want to get onto that yet. I'll come back [to it]." And they were quite pushy about going on with it and I was thinking, I suppose that is one example of having to say, "No, just shut-up for a second." . . . I suppose that is just your experience of running a discussion compared to Robert and Jake . . . I went out of the class and was quite angry. You . . . can't treat people like that. Did you feel the same – angry?
          (Kelly Interview 1: 170–230)

Kellie's anger over what occurred in the session resulted from her perceptions of inappropriate behaviour from her colleagues whose interventions prevented Robert and Jake from carrying out their plans for teaching. Kellie reported that seeing her colleagues' responses made her concerned about how she would manage her own peer teaching responsibilities. However, at the same time that this episode triggered a strong emotional response for Kellie, she was also able to articulate important aspects about teaching from her learning.

Kellie's words vividly reveal her awareness of her feelings at different times throughout the class, how these feelings were linked to the way that the class was organised for learning, and the ways in which difficult classroom situations were apprehended and responded to. Kellie clearly recognised that the way in which Robert and Jake began their teaching, introducing themselves as though the Biology methods group were school-aged students, immediately set up an atmosphere that invited corresponding school-aged misbehaviour from some colleagues, that made her feel *"uncomfortable"*. She also saw that Robert and Jake's decision to begin their teaching with an open discussion had not been helpful for directing the learning in a productive manner. Some class members openly challenged the teachers'

---

[1] Australian colloquial expression for aggressive

authority during the discussion. Kellie saw that the teachers did little to productively deal with this situation. She linked this incident to a situation that occurred later in the session, when I was teaching the group, when I had used a particular tactic to deal with a student who had attempted, quite forcefully, to redirect the topic of discussion. Kellie recognised this tactic as one possible way of dealing with a persistent student (". . . and I was thinking, I suppose that is one example of having to say, "No, just shut-up for a second."). And, on the basis of these various experiences she was reconsidering how she would structure the learning to suit the context and the learners when it was her turn to teach her peers.

Although Kellie's words reveal, to me, a powerful understanding of the teaching and learning that occurred, it is unlikely that she viewed this incident (at least at the time) as helpful for her professional growth because of the negative emotions associated with it. Kellie's strong feelings of anger and frustration about the treatment of Robert and Jake at the hands of her colleagues preoccupied her recollection of the experience. So, even though her learning was clear to me, I wondered whether Kellie was aware of what she had learnt, beyond her feelings of discomfort? If, as Korthagen (2001, p. 75) observes, establishing "a safe climate is necessary for learning to take place" then it is unlikely that Kellie, Robert or Jake gained the desired learning from this experience, at least at the time. In managing the balance between safety and challenge, there are important differences between challenge as a stimulus for learning and challenge that is too great, and becomes a threat, with the consequence that learning is limited (ibid, 2001).

One key element in transforming this situation from a threatening experience to a productively challenging one, lay with me as teacher educator. My own response to the episode included a mix of thoughts and feelings that I explained to Kelly in reply to her question (above), "Did you feel the same – angry?"

Mandi: I did feel angry [about students' behavior] and I really agonised over, should I intervene with what is going on with Robert and Jake? I was thinking, is this like when the supervising teacher intervenes with the student teacher's class and takes control? . . . I heard Adrian saying to Robert and Jake, "Just get on with it, Just get on with it". I thought that was sensible advice and I wanted them to just get on with it. And I wondered, "What is it my role to say here?"
(Kelly Interview 1: 231–238)

My response to Kellie highlights the inner conflict that I encountered in dealing with this situation. On the one hand, if I intervened in their teaching I would be undermining Robert and Jake's authority as teachers, not allowing them the opportunity to learn to deal with this situation themselves. On the other hand, by not intervening, I was allowing Robert and Jake to be exposed to hurtful behaviour from their peers. (Interestingly, I had no concerns about Adrian offering advice to Robert and Jake.) In my perception of the situation, I set up a dichotomy that left me with only two ways to act. Either I intervened or I did not.

Since I imagined this kind of situation as one that might open up new possibilities for learning, for instance, that Robert and Jake might be compelled to find new ways to deal with the events that arose, I chose not to intervene. The tension between safety and challenge then becomes highlighted as I consider differences between the extent of the risks that I was prepared for Robert and Jake to experience, and the risks I was willing to experience myself. My need for Robert and Jake to have a challenging experience led to my decision not to intervene in their teaching. At the same time, by not acting, I avoided having to find ways of dealing with this situation myself, *in situ*. My own needs for my safety overtook the students' needs for their safety. I chose a safer (and more familiar) alternative. I waited until after the peer teaching to discuss with Robert and John, and the class, what had taken place.

Korthagen (2001, p. 75) identifies that a balance between safety and challenge is achieved when there is an appropriate ". . . distance between what a student teacher is already capable of and what is required". I learnt that establishing this balance is difficult because of the considerable skill demanded from the teacher educator not only to know about the capabilities and requirements of individuals to 'estimate' this distance appropriately, but in the teacher educator possessing sufficient self-awareness to know when she is acting on her own behalf or, on behalf of the students.

A further illustration of Korthagen's ideas came via an e-mail from Lisa about the difficulties of the peer teaching experience, generally. Lisa offered her thoughts about the demands associated with achieving a balance between safety and challenge, particularly given the diverse range of individual student needs, and drew insights from these experiences to inform for her own teaching approach.

Date: Sun, 17 June 2001
From: Lisa
To: amanda.berry@education.monash.edu.au

I also think as a group we don't really like these [peer teach] sessions because they are uncomfortable. I think that it's really helpful (and horrible at times) to feel uncomfortable, but I wonder where the line is between uncomfortable and an emotion that leads to switching off? Do you sense this? These sessions are really hard. I think it is really hard for a teacher to find the line between pushing to expand boundaries and pushing over the cliff into disengaged valley. Even more so because different people are different, so just like in our class, there are some that will lose the lust for learning almost the minute they are pushed, others will thrive on it and cope with much more. For me, I think that means I will try not to push too hard, and maybe I can teach the kids to push themselves to a comfortable limit (don't ask me how though – will have to think heaps more about that one).

Negotiating Acceptable Social Boundaries

Prospective teachers in the Biology methods class struggled, not only in constructively critiquing each other's teaching, but also in discussing their understandings of aspects of Biology content. This was particularly evident in situations in which the teacher or the learner's content was put under scrutiny. In the following example, (video transcript) one of the prospective teachers, Trudy, sought to understand the meaning of the term *allele*, while Josh, the peer teacher, dealt with her questions in a way that limited opportunities for the development of both of their understandings of this concept.

Trudy: Can I ask a question? What's an allele?

Josh: An allele is the expression of, it's the physical, no, not necessarily physical . . . Do you know what phenotype is?

Trudy: Kind of.

Josh: It's a variation on a particular trait. Now let's consider hair colour . . . consider hair colour to be a trait. Everyone's got a hair colour. Any variation on that hair color is an allele.

Trudy: So is it a physical thing? Like, is it like a chromosome?

Josh: No, it's like the expression of, it's a code . . . So red hair colour is an allele. Brown hair colour is a different allele.

Trudy: So it's like the X or the Y. Male and female. Is it the X and Y? Sorry [not to understand this]! Sorry!

Josh: I haven't heard it used for male and female. It's more commonly used to refer to eye colour, number of feet. It's a different version of the same thing. So let's move on.
         (Video: Week 14)

This incident exemplifies how the 'normal rules of courteous adult behaviour' inhibited these individuals' possibilities for learning. It is clear that Trudy had a genuine question about alleles that she wished to resolve. She persisted in asking Josh to try to help her resolve it. At the same time, she was apologetic about pursuing her enquiry; she did not want to challenge Josh's authority as teacher, or seem impolite in her approach. Josh, on the other hand, did not have a good explanation to offer her and the only tactic he used to assist Trudy was to provide her with several slight variations on his original answer, before deciding to push on with the lesson.

Interestingly, no other student took up the issue, nor did Josh call on other students to assist him in developing a more helpful explanation. In this situation, Josh's decision may well have been motivated by a view of Trudy's questions as distracting to his lesson plan and to his need, as teacher, to maintain control of the situation. Consequently, the possibilities for learning inherent in this situation were left unrealized. Josh did not learn any new ways of dealing with student questions, Trudy was left with the unresolved question: *"What is an allele?"* No new growth occurred in this encounter because neither student moved out of familiar zones of behaviour. The learning was restricted by the conventional practices for behaving as a teacher and adult learner. However, as evident in earlier examples, choosing to challenge oneself to behave differently as a teacher is a risky venture with unknown outcomes. Even more risky is that teaching is a public activity, so new behaviours must be tried out (and possibly unsuccessfully at first) in front of others. This requires trust in oneself and in others that doing so is worthwhile. Many students would happily avoid disturbing or uncomfortable teaching and learning situations because their unfamiliarity with such situations leaves them feeling unsure and uncomfortable about how to deal with them (Guilfoyle et al., 1997).

There is a difficult cultural shift required in implementing approaches to learning about teaching that are based on genuinely exploring others' ideas (such as a conceptual change approach). The teacher has to be prepared to spend time exploring learners' understandings of different ideas, to believe that doing so is worthwhile and to relinquish control of the learning environment in order to work in a way that is responsive to the needs of the learner group.

Understanding More about the Relationship between Safety and Challenge

Developing one's understanding of the balance between safety and challenge is a personal, long-term process (Korthagen, 2001). As the year progressed, Lisa recognized an important shift in her thinking about the relationship between her feelings of confidence and the accompanying sense of comfort her confidence brought, and the need to continue to challenge herself as a teacher. In her first experiences of teaching, Lisa was keen to put the ideas about teaching and learning that we had discussed in Biology methods into practice. She set herself some challenging goals: to teach in ways consistent with her beliefs; to obtain feedback from her students on the effects of her teaching; and, to maintain a critically evaluative stance in reviewing her teaching efforts. The effect of trying to put all of her ideas into practice from the outset was overwhelming and, at times, undermined her confidence in her abilities as a teacher and, her feelings of comfort in the classroom. Later in the year, as her experience accumulated and she allowed herself the opportunity to relax and enjoy her teaching, she came to acknowledge that safety (in the form of confidence) was an important prerequisite for her to be able to experience challenge in ways that were helpful rather than debilitating. The importance of developing a productive balance between safety and challenge became apparent through her ongoing experiences of teacher education as she came to recognize the paradoxical situation of needing to feel both comfortable and uncomfortable in her role, in order for her to effectively develop as a teacher. In her second interview, Lisa captured her thinking about this issue:

> Lisa:　I think I just relaxed a lot and it was much easier and the thing that was the most challenging was trying to push myself to be uncomfortable because I was enjoying it and enjoying feeling a bit more relaxed and a bit more comfortable and I didn't want what happened on the first round to happen on the second round . . . to lose confidence. I could just see the effect that my confidence had on the students in a positive way and I didn't really didn't want to lose that for myself and also for them too . . . I think the challenge was

and it will be next year as well I reckon is to find a balance between comfort and . . . like I needed to feel some level of comfort to be a good teacher but I wanted to make myself uncomfortable to be a better teacher.
(Lisa Interview 2: 10–13)

Lisa recognised the impact of the teacher's feelings of confidence on the learner's confidence in the teacher. As her feelings of confidence and hence competence grew as a teacher, she noted a corresponding positive effect on her students. Her understanding of the interdependent relationship between learner and teacher growth also led her to make some choices about how she interacted with me.

SEEING THIS TENSION THROUGH 'ANOTHER'S EYES'

So far, I have focused on an examination of the tension between safety and challenge in terms of the ways in which *I* negotiated the balance between challenging prospective teachers and being hurtful towards them. However, it was not only me who felt this tension. Through discussions of our teaching, Lisa helped me recognize that prospective teachers also experienced this tension in their feelings towards me. For instance, in offering me feedback about my teaching Lisa worried to what extent the things that she said/wrote to me were helpful and what was hurtful. Although we shared a belief in the value of honesty in our interactions, Lisa decided that sometimes too much honesty could be hurtful, and because there were some things that may have been too uncomfortable for me to hear, she decided not to risk telling me. An e-mail to me explains her ideas:

Subject: DISCOMFORT
Date: Sat, 16 June 2001
From: Lisa
To: amanda.berry@education.monash.edu.au

Although I'd like to think that I am pretty honest, I have to admit that I sometimes hold things back if I think they will really offend her [Mandi]. Mandi has really worked to establish an environment in which it is OK to make honest comments about our learning and has really encouraged us to feed back to her. Despite this, I still hold back sometimes. But . . . does this really matter? Maybe it's a good thing. I'm a person as well as a student, so I guess I have some ability to judge what is honest and helpful and what is honest but possibly hurtful. I use my judgment to decide how much 'uncomfortable' information to give to Mandi . . . and maybe that's OK, because there must be a point when honest feedback that is hurtful becomes so horrible to receive that it's not helpful anymore. So perhaps if my students are the same, and hold back honest feedback [to me] because they don't want to hurt my feelings, I will still get some great, helpful information. After all, I probably can't deal with all the information they give me anyway, so if they selectively keep some more 'uncomfortable' information to themselves for a while, I think that might be OK.

Lisa's words powerfully illustrate the interpersonal dimensions of the tension between safety and challenge. Her consideration of the issue of *"what is helpful and honest compared to what is hurtful and honest"* in her feedback to me highlights her understanding of teaching as a personal experience. The feelings that emerged for Lisa as, *"a person as well as a student"* led her to make decisions that were based not only on the ways in which the feedback she offered might impact the cognitive aspects of my learning about my practice, but also in their social and emotional impact on me. In coming to understand her experiences, including how she might act towards me, Lisa considered the situation from the point of view of working with her own students, including their likely needs and responses. Lisa's response is a good example of the way in which I hoped that prospective teachers' experiences of my teaching might prompt their thinking about their approaches to teaching their students.

Building and Risking Relationships

The development of productive personal relationships requires knowledge of oneself; and of oneself in relation to others. My intent was to build these prospective teachers' confidence in themselves and their ideas as well as aiming to extend their view of practice, so that through growing confidence, they would be encouraged to push ahead, not simply remain comfortable with their existing practice. One of the continuing struggles that I faced in implementing an approach to teaching about teaching that aimed to challenge and confront prospective teachers' views of learning to teach was my fear that in so doing, I would jeopardise my relationships with them.

As a person who defined herself in terms of her relationships with others, I found it particularly difficult to teach in ways that aimed to disturb prospective teachers' thinking about their pedagogy. My previous experiences of implementing this approach in other subjects (3rd year, B.Ed.) were based on partner teaching with a colleague, so that stepping out to risk new practices could be supported and discussed with another teacher educator and the class. In Biology methods classes I was working alone. This led me to feel more vulnerable and sensitive to the responses of the prospective teachers.

Fear of compromising my relationships with the students sometimes prevented me from acting in ways that might have pushed them a little harder to consider the reasoning behind, or effects of, their actions as teachers or learners. At other times I worried that the actions I took may have upset the relations I had established with particular class members. Other teacher educators have also experienced these feelings as they have attempted to push their students to consider hard questions about teaching and learning. For example, Schulte (2001) noted that "Engaging students in this kind of confrontational pedagogy [is] a challenge for me, because my self-identity is often closely tied to my ability to relate to others" (Shulte, 2001, p. 7). Hence, choosing to act in a way that might jeopardize the relationships so important to teaching may well (rightly) be too great a risk for many teacher educators.

SUMMARY: WHAT DID I LEARN FROM EXAMINING THIS TENSION WITHIN MY PRACTICE?

In traditional approaches to teacher education, avoiding uncomfortable situations actually diminishes the possibilities for learning and often, such avoidance is due to a lack of the very trust, confidence and sense of relationship that is so important in teaching and learning about teaching. For me, being able to recognize and/or create potential learning situations that challenged others to reconsider their ideas about teaching was demanding and idiosyncratic. Some kind of learning intervention though needs to be explicit if genuine progress in learning about teaching is to occur as helping prospective teachers 'feel what it is like' to be in a position in which they do not know how to respond is an important first step in learning about practice. Yet, to act in this way entails risk.

In my readiness to create a context for prospective teachers' learning that attempted to push them beyond the comfortable and familiar, I forgot, or neglected to acknowledge, the important role of feelings in teaching. Instead, my own desire to challenge the familiar ways of working that these students brought with them, often overwhelmed my ability to recognise and respond appropriately to their individual needs, including their feelings. I wanted to develop a pedagogy that sufficiently disturbed prospective teachers' thinking about teaching that they had to consider alternatives to the comfortable and familiar. But it was difficult for me to know how far I could go before the disturbance intended to initiate learning actually prevented it.

Obviously, risk taking was real and different for all involved. The degree of risk varies greatly from individual to individual and finding optimum value through risk taking is itself risky business. Choosing to act in ways that challenge traditional notions of maintaining the status quo is both emotionally and pedagogically challenging. In essence, the tension between safety and challenge as it played out in my practice, illustrates that, as a teacher educator, I needed to:

- know enough about what was likely to be uppermost in prospective teachers' minds (i.e., their needs and concerns)
- know my own goals for prospective teachers' learning (i.e., where am I trying to move them towards)
- listen carefully to what prospective teachers say such that I could work out when there is more than the face value message being expressed (e.g., asking myself, "What messages do I really need to pick up on here?")
- know each student sufficiently (to consider what risk might be acceptable for that person).

In addition, I needed to know about the selected Biology pedagogy so that I could help explore, challenge and support the development of prospective teachers' Biology knowledge. It is little wonder then, that the process of learning to recognize and deal with this variety of factors was a challenging, complex and confusing process for me.

Chapter Nine

PLANNING AND BEING RESPONSIVE

plan (n) formulated and especially detailed method by which a thing is to be done; an intention or proposed proceeding.

responsive (adj) responding readily to some influence, appeals, efforts, etc.; sympathetic.

> Be confident to be responsive to possibilities in learning experiences. (Berry & Loughran, 2002, p. 19)

INTRODUCTION

This area of tension, between *planning and being responsive*, focuses on the interaction between teacher educators' planning for their students' learning and responding to learning opportunities as they arise in practice. Such learning opportunities or, unplanned "teachable moments" (van Manen, 1991; Hoban & Ferry, 2001) become available when prospective teachers' needs and concerns can be recognized and responded to 'on-the-spot' within the learning context. However, since such moments cannot be planned in advance, considerable teacher educator expertise is required to be sensitive to a potentially "teachable moment" and to make a decision about how, or whether, to respond.

The ways in which I came to recognize, and then manage, the competing forces of teaching with a predetermined plan and being responsive to "teachable" occasions that arose within the context of the Biology methods class are illustrated through the tension described in this chapter. First, I explain how I organized Biology methods so as to create an environment that would support teaching about teaching in a responsive manner. Then, I examine the factors that affected my ability to recognize and respond to unplanned learning opportunities within Biology methods sessions. Finally, I identify the learning about practice that occurred for my students and me as a consequence of exploring this tension within this research.

BUILDING A RESPONSIVE ENVIRONMENT

In developing the Biology methods curriculum I strived to create an environment that encouraged risk-taking and genuine experimentation in teaching so that as individuals, and as a group, we might challenge and extend our understanding of our teaching. I anticipated that this approach would support all participants to begin to develop personally and pedagogically meaningful approaches to practice. My ideas about teaching Biology methods are based on a view of learning as a shared responsibility of all (i.e., teacher educator and prospective teachers). While I had identified broad goals and purposes for the Biology methods curriculum, I believed that as a group we would collectively identify more specific, personal goals as the year unfolded and as these prospective teachers' understandings of their needs developed. Therefore, in my approach to preparing methods sessions, rather than strictly prescribing and controlling the learning experiences, I attempted to create conditions for learning (Loughran & Northfield, 1996), so that aspects of learning about teaching appropriate for the group (or particular individuals) could be highlighted as they arose. To do this, I focused on providing an environment rich in learning experiences (e.g., Munby & Russell, 1994). From an experience rich context, incidents and questions that raised opportunities for learning about teaching could be capitalized on, and explored, together.

Creating such an environment required both attitudinal and organizational shifts: prospective teachers needed to believe that such an approach was worthwhile (compared with me giving them techniques for how to teach) and I needed to organize Biology methods sessions in such a way that opportunities for learning about teaching (both planned and unplanned) could be raised and pursued. Planned activities included for example, peer teaching, whereby class members were encouraged to take a risk to try out an unfamiliar approach to teaching a particular Biology topic; unplanned activities resulted from spontaneously arising situations within a session that I recognized as potentially powerful contexts for exploring teaching and learning. However, knowing what I wanted to do and knowing the reasons why I thought such an approach to learning about teaching was powerful and valuable did not adequately prepare me (or these prospective teachers) for some of the difficulties we encountered in implementing this pedagogical approach. These difficulties were borne partly out of my expectations of my competence in working this way, (I expected to be able to regularly engage all students in powerful learning experiences based on needs arising within sessions) and partly, from these prospective teachers' expectations of learning to teach Biology. (My approach was perhaps perceived as unusual and perhaps not always helpful, i.e., not one that helped them to learn how to teach Biology.) As a consequence I often felt frustrated and unconfident about how I could effectively enact these ideas well, in Biology methods classes.

Noticing Possibilities for Learning

In viewing the videotape after the first (recorded) Biology session I noticed that one of my students raised an issue that several others also seemed keen to discuss. At the time, my response to this 'unplanned' discussion was to finish it quickly in order to proceed with my intentions for the session. Watching the video made me conscious of how my predetermined agenda influenced my response to this situation. While the actual topic of discussion may or may not have been helpful for the class to explore further at that time, what struck me from viewing this brief incident was that this was an example of a situation where I was faced with alternative possibilities for action. The possibilities included following the discussion or, continuing with my plan. In this case, my concern about the (limited) amount of available time led me to stop the discussion and continue with my pre-planned agenda.

Two weeks later a similar situation arose. However, in contrast with the preceding example, this time I chose to follow the needs of the students, rather than impose my predetermined plan. Interestingly, when I watched the videotape of that day's method session, I wondered about the accuracy of the 'indicators' that I had used to inform my decision to continue the discussion.

> We talked more about the camp at the beginning of this session [than I intended] . . . I was getting the feeling it was right to continue with a longer discussion of camp, because those who hadn't attended seemed interested to know about the experience and those who attended seemed quite keen to discuss some examples of [school] students' behaviour. (Was that a good judgment on my part or was that only a couple of people who were pushing the discussion along and others were just politely continuing?) (Week 3, Open Journal)

In this case, my perception that it felt "right" to continue the discussion was based on the opinions of several of the more vocal students, a fact I was not aware of until I viewed the videotape, later.

While these two examples are relatively minor situations of choosing between a predetermined action and being responsive to unplanned events, they are nevertheless helpful in illustrating general characteristics of the tension described in this chapter as it operated across a range of different situations I encountered. To begin, both situations illustrate that within the environment of a teaching situation the teacher/educator perceives a variety of different signals, or stimuli, that alert her/him to pay attention to particular features of that environment. The kinds of signals that the teacher educator pays attention to are determined by the particular sensitivities that she/he brings to the situation, including knowledge of learners, self and context. Mason (2002) identified that an individual's choice to act in a given situation is governed by two factors: awareness of "a possibility to choose, that is, recognizing some particular situation about to unfold" (p. 72); and, having alternative actions or behaviours to choose from, relevant to that situation. The first factor, awareness of choice, involves being consciously alert to possibilities for action in a situation, compared with teaching in habitual, or 'unthinking' ways. For instance, in the second example (above), my awareness of what was happening within the class at that time led me to encourage the discussion to continue.

Noticing possibilities for action may occur immediately, within the moment of teaching, or later, after the event. The first example illustrates a small instance of retrospective noticing. The videotape prompted my recognition of a moment of choice that was perhaps not apparent to me at the time of teaching.

The extent of a person's conscious awareness of being in a particular situation varies from being vaguely aware of the nature of events that are occurring, through to being fully aware and taking action in response to those events (Mason, 2002). One of the continuing struggles that I faced in teaching Biology methods and a source of the tension described here, was that I strived to be consciously aware, at all times, of all that was happening within a session so that I might not miss an opportunity for capitalizing on learning about teaching. The problem was that I saw "teachable moments", everywhere. I struggled to know how to deal with the myriad of available possibilities. Two brief journal extracts below, illustrate this point.

> There are so many alternative possibilities to be explored!
> (Personal Journal, Week 1)

> I'm so conscious . . . as students are doing something . . . I am thinking, what am I doing? What are they doing? What am I looking for?
> (Personal Journal, Week 12)

Interestingly, I was often so busy trying to identify "teachable moments", that I actually missed some of the most powerful unplanned opportunities for learning, a point I will return to later in this chapter.

At the same time that I recognised possibilities for learning about teaching everywhere, I often found it difficult to know how I might act in order to highlight the issues that I saw in ways that were helpful and meaningful for these prospective teachers. This problem links to the second factor identified by Mason (2002) governing an individual's choice to act; that is, having (and using) appropriate behaviours to bring about possibilities in a particular situation. There were many instances in Biology methods sessions where I found myself recognizing a situation that presented possibilities for new learning, yet that I did not act upon. This was because I was unsure how to act in a way that could effectively highlight the relevant issues, or because I felt somehow 'paralysed' and unable to act, other than letting the flow of events roll forward. For instance, in Week 1, I noticed that one of the class members, Andy, approached a group work task in a way that showed his routines of learnt behaviour.

I thought that Andy's behaviour was worth highlighting for the group because I knew that prospective teachers were often unaware that they had developed routine ways of acting in response to certain tasks or situations, but that they expected their students to know and use these routines. At the time, however, I did not act on this moment because I felt unsure of how to do so appropriately. I wrote about this incident in my personal journal as I watched the videotape after that class.

> Andy takes the information sheet and straight away moves his chair to be part of the group. How do I pick up on this in a way that is useful to talk about? It's something that kids in a classroom wouldn't do, but that he does automatically. I want to alert them to assumptions that they will make about what students will do, because they do it. I could have said, "Andy, I noticed you moved your chair . . ."
> (Personal Journal, Week 1)

Although I could have simply told the prospective teachers about what I saw and how that might serve as a problem for them in their teaching, I knew that telling them was less likely to impact their thinking than if they were able to somehow experience the consequences of their actions. Hence, this moment passed because I could not think of a way of setting up a situation that would help highlight the assumptions embedded in Andy's approach.

In another example, my uncertainty about how to act in response to a particular situation was influenced by my concern not to upset one of the prospective teachers, by my choice of actions.

> Josh [student teacher] went through the chapter and made sense of it for himself, then turned the chapter into a diagram and then expected that we could follow his thinking. A number of us couldn't and he couldn't understand why. Apart from telling him this, and I don't think that this would work anyway, and he would probably be offended, how do I deal with this? How do I create something out of this that we can all discuss thoughtfully?
> (Personal Journal, Week 14)

My Personal Journal contains numerous self-admonishments over missed opportunities for prospective teachers' learning. I constantly grappled with the dilemma of wanting to work in a responsive manner yet, at the same time, feeling as though I did not have the confidence or skill repertoire to do so effectively. Occasionally, I was able to resolve my difficulties. For instance:

> Today issues like the difficulties of translating ideas from textbooks came out because I felt free to let them come out.
> (Personal Journal, Week 3)

Often, the more I tried to work at being consciously aware of what was happening in Biology methods classes, the less I seemed able to act in a responsive manner. When I allowed myself to relax, to be less vigilant in looking for opportunities for learning about teaching, I was better able to capitalize on experiences for learning when they arose. This presented me with an ongoing dilemma and, for me, lies at the heart of the tension between planning and being responsive. To be responsive I needed to be open to what was happening around me, yet in trying to be open, I felt confused and overwhelmed by having so many possibilities to choose from.

MAKING MYSELF A "MODEL OF DIFFICULTY"

Sometimes I made my dilemmas explicit to the methods class. I recognized that the feelings I experienced in these situations were not dissimilar to those that they would likely experience as they grappled with decisions about when and how they might suspend their lesson plans to capitalize on an opportunity for learning with their students. I also hoped that by giving access to my decision making processes and difficulties that these prospective teachers might begin to consider more closely the kinds of factors that influence a teacher's ability to be responsive, and perhaps learn to recognize and articulate some of these moments in their own teaching. Perhaps too, if they saw me struggle with decisions about my teaching then they may feel more reassured that such feelings were considered normal. One such occasion of sharing a dilemma with the methods class occurred early in the year.

The students were working on a series of tasks based around learning about the human immune system. At the beginning of the session I distributed a diagnostic tool designed to investigate their ideas about how the human body defends itself against disease, which I wanted them to revisit at the end of the session. The purpose of the task was to provide prospective teachers with an experience of a diagnostic tool that they might consider using with their students, and, at the same time, prompt them to consider the extent of their content knowledge about this topic. As the session drew to a close, I realised that it was time to revisit the diagnostic task, yet they were still busily engaged in another activity. As I felt the dilemma of deciding whether to stop their current activity and return to the diagnostic task, so I told them about my feelings. Because I knew that this would be a situation that they would also regularly face in their teaching I wanted to make explicit my thought processes and to use this experience as a learning opportunity for all of us.

> Mandi: I'm in a dilemma. My difficulty is that I can see that you're getting in
>   to the task and that is not a good time to stop. However, since one of
>   my purposes today is to get you acquainted with a range of different
>   approaches to knowing about the human immune system, then I think
>   it is important to ask you to stop what you are doing now and go back
>   to the sheet you had at the beginning of the lesson, to write down any
>   new things you've learnt and any questions or puzzles that have come
>   up for you, or that you may still have.
>   (Video: Week 4.)

In sharing my thinking with the class, I highlighted my recognition of a particular kind of situation (running out of time in a lesson), the choices I had available (I could keep going, or revisit the initial activity) and the reasoning behind my decision (the purpose of the session was to experience a range of approaches to learning). In making my thinking explicit, I anticipated that these prospective teachers might learn to recognize some of the complexities associated with planning for a particular experience and working responsively with a plan as it is enacted. I wanted to let them know the purpose behind my actions, so that they might be more inclined to consider the purposes behind theirs.

My familiarity with this type of regularly occurring teaching situation (i.e., where an activity takes longer than the teacher anticipates) led me to recognize it when it arose, to feel comfortable about how I might deal with it, to deal with it publicly and to use it as a context for learning about teaching. In other words, I felt confident to make use of myself as a "model of difficulty" (Mason, 2002, p. 145) in this situation in order to promote prospective teachers' learning. However, in other situations, where unexpected or unfamiliar situations arose, and I was genuinely struggling to know how to act in response to a situation, it was much harder for me to discuss this with students. In such instances, my worries over how I would manage the situation usually precluded any thoughts of sharing my concerns with the class. One example occurred during a peer teaching session during which several prospective teachers had voiced criticisms of their peers as teachers. I felt concerned that they had acted this way towards each other, particularly when their behaviour towards me, as teacher, was quite different. I wanted to draw their attention to these differences but in the 'heat of the moment' I did not, possibly because I was too strongly immersed in the emotions associated with the experience.

> I told everyone in the Biol method group that this was a safe place to try things out, to experiment and to risk take [but] then . . . I let people who were teaching be exposed to behaviour from their peers that was threatening and rude and I didn't do anything to stop it . . . Because I am reasonably relaxed about having my teaching publicly debriefed, then I didn't recognise what it must have been like for those who are inexperienced at it, and who hadn't chosen to do it but it was part of their requirements for this subject.
>
> After the peer teaching was finished I ran a learning activity with the students and I noticed a massive difference in the behaviour of the group. They behaved so differently for me than they did for their peers. And, even though I noticed this when I was teaching, I didn't think this out loud [make explicit] to them. I don't know why, but I just couldn't. Something simple like 'hang on a moment the dynamic in here has really changed. What's going on?' perhaps would have been a start. (Personal Journal, Week 13)

In hindsight, I believe that highlighting the problems that I was experiencing might have helped these prospective teachers to recognize the realities of teaching – events don't always work out as the teacher intends, and sometimes the consequences of one's actions can be very uncomfortable. I might have helped them to recognize the possibilities inherent in such a situation (as well as the problems) by asking them if it had been them in that situation, how might they have acted? In such a way, they might have been helped to understand the problematic nature of the situation and possibly increase their repertoire of behaviours in learning to deal with similar situations in their future teaching. The problem was that my anxiety about how to deal with the situation restricted my capacity to act in a responsive manner. Instead, I did what many prospective teachers often do in similar circumstances: I aimed just to get through the session. As a consequence, a genuine opportunity for learning about teaching was lost.

This episode also highlights an aspect of my understanding of the nature of responsiveness. Being responsive meant responding to a situation as it happened and, 'on the spot'. I did not consider the possibility that the issues raised might have been revisited and discussed with the students in the following (or later) sessions, when presumably I would have felt more emotionally detached and hence better able to explore these issues in productive ways. A sense of urgency guided my thinking that seemed to block possibilities for acting within the moment and preclude possibilities for returning to them, later.

SETTING UP CONTEXTS FOR LEARNING AND RECOGNIZING POSSIBILITIES WITHIN THEM

Sometimes I deliberately manipulated situations in order to bring to the surface particular aspects of teaching and learning that I thought were worthwhile to bring to the prospective teachers' attention. Peer teaching was an ideal context through which I could highlight for prospective teachers some of the more complex dimensions of learning about teaching from what was seen or felt by individuals as they interacted with their peers. Deciding what to highlight for learning was something that I could not plan in advance since situations needed to be apprehended within the immediacy of practice as they emerged from interactions between different individuals (including me), the context and the content of the session. However, there were some 'typical' situations that I could predict were likely to emerge as prospective teachers employed particular approaches to teaching, such as group work, chalk and talk, discussion, etc.

Features of the tension between planning for learning and responding to learning opportunities in practice, became evident to me as I attempted to direct particular peer teaching situations in a specific manner because of the possibilities for learning that I recognized within them. As mentioned earlier in this chapter, in my eagerness to pick up on teachable moments and direct the learning in ways that I saw as helpful and meaningful, I sometimes failed to notice some of the most powerful unplanned opportunities for learning.

A compelling example of this occurred during Bill and Joanne's peer teaching when I attempted to highlight for the class the ways in which particular rules for group interactions influenced the nature of individuals' learning in a group. (This episode is explored in detail in an earlier chapter hence I revisit this incident only briefly here.) Through making this incident into part of the learning, I hoped that Bill, Joanne and the rest of the class might be prompted to look more closely at how they constructed group work experiences in the future. It was not until later, when I shared a transcript of this session with a colleague that I came to recognize that I had overlooked other aspects of the group interactions that day because they had not been part of my agenda for the class.

Discussion of this episode between my colleague and me helped me to focus on new aspects of the session. We shared a transcript of the video, along with some notes I had made while watching it.

John: (Reads from transcript notes) "Bill, head down, seems to be trying to avoid eye contact." So did you notice that at the time, or only afterwards?

Mandi: Only on the video.

John: Why not at the time?

Mandi: Because I was busy talking to Joanne and Natalie and I had an idea in my head . . . and I can't remember registering it [Bill's body language].

John: Here's one of those dilemmas that you face all the time. There are things that are happening at different levels and you're trying to work out all the time which level to play at for a while to make it meaningful for the student . . . Because if you had spotted that, then there are a whole lot of other things you might have said to yourself like, "I wonder why he's sitting like that? Hmm, maybe does he feel like his actions or ideas have been dismissed?" Not that you can really answer that but all those things are happening to you . . . It's a matter of making a decision. In this case, you might say, I'm going to drop all those other things that are of interest to me because I think I recognise something that I have done or this [experience] has done that has now impacted on him and this could be a very powerful session. So you ask, "What's the problem Bill?" Or, you say to yourself, this is not the time to do it.
    (Colleague discussion, 21/06/2001, 150–173)

At the time of the class I did not notice Bill's body language. Instead, I was preoccupied with encouraging the prospective teachers to recognize, and talk about, aspects of their experience that I had thought important, i.e., problems associated with group work. I did not pay attention to Bill's response in the way I paid attention to my own agenda. In fact, I did not consider Bill's response as something that *could* have been part of the learning. It was only later, when I watched the video and discussed it with my colleague that I began to understand more about the ways in which my perceptions of the situation influenced what I was able to see and what I wanted these prospective teachers to see.

This situation makes more complex the notion of being responsive as a teacher educator: there is a difference between responding to what one thinks is happening in a particular situation and *really tuning in to the specific needs and concerns of individual students*. Genuine responsiveness involves paying close attention to one's students, beyond what the teacher educator expects to see or hear. Other teacher educators have identified the importance of carefully listening to one's students in order to act responsively towards them. For example, McNiff, (1996), a teacher educator who studied her practice, learnt that listening is a complex act that requires a sensitive appreciation of the views of others.

I am a better teacher now than I used to be, because I listen. I have discovered that listening is not just a physiological activity but something that is undertaken holistically. It is a mixture of gesture, inspired guesswork, experience, and striving to be at one with the other. It is a business of reaching out. (p. 3)

Another teacher educator, Northfield, learnt from his experiences of teaching high school students that teacher confidence, experience and the ability to listen to students were important factors influencing the success of unplanned teachable moments (Loughran & Northfield, 1996). Northfield wrote, "My successes . . . came largely from unplanned opportunities when I listened to students and had the confidence and experience to respond at the time" (p. 138). In some ways Northfield identifies a paradox for the teacher/educator since confidence to act is derived from the experience of acting, so one must act in order to build up knowledge of how to act, even though one may be unsure of how to act at the time. Experience then both precedes and informs understanding of practice; a point that was very difficult for me to accept and 'live' as a teacher educator. These ideas also link with the experiences of prospective teachers as they begin teaching their students when they have little experience and often little confidence, and hence underscores the importance of teacher educators making such issues explicit and learning about them, together with their students.

Prospective Teachers' Learning about Learning to Respond

Through teaching about teaching in a way that aimed to be responsive to the needs and concerns of prospective teachers in the Biology methods class, I hoped that they might be prompted to consider their approaches to teaching as a responsive activity, with their students. Evidence that at least some of them were beginning to recognize such issues for themselves emerged through interview, although their thinking seemed to have been more strongly influenced by their school-based experiences than their experiences of Biology methods classes. For example, Tina reported that her practicum supervisors helped her to recognize that her predetermined expectations for her students' learning restricted opportunities for supporting students' learning for themselves. Tina said about her own practice:

Tina: Something my supervisors said to me when I did a concept map was, "Well you went in there and had your own idea of the concept map and these kids they gave you the answers and you only put down the ones that fitted your concept map. So you created your own concept map and gave it to them rather than them thinking about their learning and putting it into their concept map."
(Tina Interview 1: 54–55)

Tina's practicum experiences highlighted her need to be sensitive to understanding a learning situation from the point of view of the learner, rather than imposing predetermined teacher ideas about what might be learnt, and how.

Another prospective teacher, Jacqui, recognized that feeling comfortable in the teacher role and knowing about her students as individuals were important factors contributing to a teacher's ability to genuinely listen and respond to her students, compared with 'simply' following a routine.

Jacqui: In the discussion situation when the kids answer you, you don't just go, "OK good, next question" . . . I think if you are more relaxed you are more open to thinking about what they [students] are saying . . . You've also got to be relaxed with the class, not just the teaching . . . Maybe if you got to know them on a more personal level . . . you would feel more confident with them.
(Jacqui Interview 1: 29–31)

There are strong parallels between the kinds of experiences reported by these prospective teachers and my own experiences of learning to teach about teaching in a responsive manner. Knowing about students as individuals, feeling comfortable in the role of teacher/educator and looking beyond one's predetermined expectations of a learning experience were equally important for me to take into account as they were for these prospective teachers. Such ideas are echoed in the writing of Dewey (1933) who identified that teachers' possession of particular attitudes and types of knowledge grows confidence to be open to new possibilities for learning.

> Flexibility, ability to take advantage of unexpected incidents and questions, depends on the teacher's coming to the subject with a freshness and fullness of interest and knowledge. There are questions that [the teacher] should ask before the recitation commences. What do the minds of pupils bring to the topic from their previous experience and study? How can I help them make connections? What need, even if unrecognized by them, will furnish a leverage by which to move their minds in the desired direction? (p. 276)

Understood within the frame of teacher education, Dewey's ideas focus on knowledge about prospective teachers' prior beliefs and experiences (of teaching, learning, and in this case, Biology), what the teacher educator wants the prospective teachers to learn (about teaching, learning and Biology) at a particular time, and what the teacher educator recognizes that prospective teachers need to know that they do not yet know about these fields, and ways to move them towards this knowledge. The central reason why the teacher must have "abundant knowledge" in these areas (and that is not often recognized for its importance) is because: "The teacher must have his [sic] mind free to observe the mental responses and movements of the student[s] . . . The problem of the pupils is found in *subject matter*; the problem of teachers is *what the minds of pupils are doing with this subject matter [italics in original]"(p. 275)*. Viewed through the lens of teacher education, this means that until the teacher educator has accumulated sufficient professional knowledge of teaching about teaching she will struggle to listen to the needs, concerns and responses of the prospective teachers. The professional knowledge required is knowledge about teaching about teaching generally, and knowledge that is specific to working with a particular group of students at a particular time. The complexity of all of this needs to be acknowledged and understood in order to incorporate it into the practice of teaching about teaching. Hence, this offers some insight into what seems to be a contradiction between the opening quote for this chapter, *"Be confident to be responsive to possibilities in learning experiences"* and my hesitation in responding to learning opportunities that I encountered in Biology methods.

While confidence is a necessary teacher educator attribute, confidence alone is not sufficient to support teacher educators' abilities to act responsively. My confidence to act in a responsive manner was tied to having adequate knowledge of the needs of these students and the specific demands of this subject. Responsiveness is difficult and not necessarily easily learnt, as one needs to develop an attitude toward it being of value in order to learn to want to look for it in practice.

TEACHER EDUCATOR EXPERTISE: A COMBINATION OF FACTORS

Through the study of my practice I have come to learn more about the ways in which I perceived and responded to different planned and unplanned events within Biology methods classes. From examining different instances from within my practice I have come to recognize a range of factors that influenced my capacity to notice and respond to these events, and more generally, the demands that being responsive placed on me as teacher educator. Factors I have identified included:

- the capacity to sensitively tune in to the needs of individual students (even when students may be unaware of, or unable, to articulate these needs themselves);
- an ability to distinguish between different kinds of learning opportunities (including distinguishing between my needs for prospective teachers' learning and prospective teachers' needs for their own learning); and,
- a genuine preparedness to share intellectual control with prospective teachers.

As these different factors are developed and better understood within the context of teaching about teaching, so I would argue, professional knowledge of teacher education is developed.

Teachable moments vary according to the particular needs and concerns of individual prospective teachers. However, embedded within the complexity of a teachable moment lies the teacher educator's ability to discern between his/her own needs and concerns and those of the prospective teachers. Recognizing differences between one's own needs and those of less experienced others, and deciding which to respond to is neither simple nor straightforward. And, inherent within this complexity is the difficulty that recognising a teachable moment is separate and distinct from being able to respond appropriately to it. At the heart of the issue is the teacher educator's role in, ". . . focusing the attention: some things in the environment become important, others are disregarded: what is important is what helps to satisfy the need" (Korthagen & Lagerwerf, 1996, p. 165).

SUMMARY: WHAT DID I LEARN FROM EXAMINING THIS TENSION WITHIN MY PRACTICE?

The elements comprising the tension between planning and being responsive are embedded in the development of an informed understanding of one's practice as a teacher educator. As a consequence of examining this tension within my practice I have learnt that the capacity to act in a responsive manner requires:

- *Working on different levels simultaneously.* This means identifying prospective teachers' needs within a teaching situation and being able to distinguish between what is relevant and meaningful to highlight at that time, compared with what might be raised and revisited at a later stage. Some of the difficulties I experienced resulted from my enthusiasm to raise learning opportunities from every possibility I saw, rather than selectively exploring situations. This links to the next point,

- *Knowing what is important in the 'big picture' of prospective teachers' learning about teaching* and being guided by big picture goals rather than worrying about immediate, short term plans. In my approach to teaching, I struggled to deal with situations that presented "teachable moments" because I thought that this would somehow compromise my intended plans for students' learning within a session. I often felt as though I had to abandon the plan in order to be responsive, rather than recognising that my plan did not have to be wholly abandoned, just steered a little differently, while working towards the same goals. My thinking was quite rigid and dualistic in this respect, as I considered only 'either–or' possibilities (i.e., to keep going with my predetermined plan, or respond to this situation). I needed to keep in mind a stronger sense of the broad goals I was working towards so that as situations arose within sessions, I could consider their potential contribution to these ends, rather than viewing them as distractions.
- *Trusting oneself and having confidence to follow through* in a situation that requires risk. Making decisions to pursue new possibilities entails a great deal of uncertainty and personal risk for the teacher educator, since one must be prepared to encounter the unexpected (Britzman, 1991). In these situations confidence matters, including the confidence to act upon a teachable moment and to do so in a manner that encourages student teachers to feel willing to engage in this new experience, too. Even though I understood the idea of uncertainty as a constant feature of teaching and learning about teaching, in my actions I often sought certainty as I looked for ways to know that I was making the best choices for supporting student teachers' learning about teaching. My colleague reminded me that having faith in myself to act was important.

Mandi: It's possible to see learning opportunities in everything, so how do you become selective? How do you know what's good to pick up on?

John: You have to ask yourself, is it worth them hearing this? Sometimes you don't know. Like Schön, you've got to step out in faith sometimes.
(Colleague discussion 15/06/01: 120–130)

The need for certainty manifests itself in attempts to control the learning experience. At the same time that I expressed an intention to work flexibly, in a way that acknowledged and built upon prospective teachers' needs, I often held quite tightly to the planned experiences for a session. Knowing that I acted in this way was not enough to change my capacity to act differently at that time.

- *Recognising differences between reacting and responding.* There is a difference between simply reacting to events as they occur and responding thoughtfully on the basis of what one knows about the experience of learning to teach. In reacting to a situation, the teacher educator behaves in an unthinking manner – one that is not based on the particulars of that person (or people) in that context at that time. Responding requires a basis of understanding when one is making a choice about a situation, why and possible alternative behaviours relevant to that situation, as well as some ideas about student teachers' likely responses. Although I had a reasonable grasp of the relevant theoretical knowledge associated with learning to teach, bringing this knowledge to mind, and linking it to situations which were occurring in the methods class was difficult because of my limited experience of the particular types of situations I encountered. Responding also involves distancing oneself from the emotions associated with an experience; reacting is emotionally led (Mason, 2002).

By taking hold of teachable moments prospective teachers may become aware of aspects of their environment that they could not previously see. Teaching educator openness to seeing these moments, feeling confident to respond and having a reason to feel confident comes from familiarity with particular features of experience as well as theories about teaching and learning that can be used to link the propositional and experiential knowledge together meaningfully for prospective teachers. Also, it comes from knowing about one's students individually, including their needs and concerns. And, it includes recognizing differences between one's own needs and concerns and how these drive one's agenda in ways that may or may not support prospective teachers' learning.

Chapter Ten

VALUING AND RECONSTRUCTING EXPERIENCE

Experience (n) actual observation of or practical acquaintance with facts or events; knowledge or skill resulting from this.

> Experience cannot be taught, it must be had . . . There is something you need to know, but your teachers cannot tell you what it is. (Munby & Russell, 1995, p. 175)

INTRODUCTION

Prospective teachers' prior experiences serve as powerful templates for the ways in which they think and act as teachers (Korthagen, 2001; Knowles & Holt-Reynolds, 1991). These prior experiences strongly influence prospective teachers' expectations of their preservice programs, reinforced by popular stereotypes about teachers' work (Britzman, 1991). While their experiences of 'seeing' teaching are extensive, few prospective teachers have actually experienced the 'doing' role of teacher, a situation that results in a curious paradox: teaching is at the same time very familiar to prospective teachers (from years of observing teachers at work) and yet, unfamiliar (many have not actually taught in a classroom before). Their familiarity makes them well prepared to make sense of teaching, since: "They [prospective teachers] already possess quantities of experienced-based information on virtually every topic or concept we plan to teach" (Holt-Reynolds, 2004, p. 346), but also strongly limits their sense making processes, because of ingrained ideas about: that which is good teaching; learner capabilities; and, what works in the classroom.

The tension described in this chapter emerges from my teacher educator role in helping prospective Biology teachers to recognise the value of personal experience in learning to teach (including those experiences they bring and those they have in their teacher preparation), and helping them to see that there is more to teaching than experience alone.

Moving beyond an 'experience alone' view of learning to teach happens when the ideas and experiences that prospective teachers bring to, and gain through, their teacher education are explicitly acknowledged and valued, and are used as a basis for developing new understandings of practice that they may not yet have considered. The pedagogical challenge embedded in this for me as a teacher educator, and hence the source of tension between *valuing and reconstructing experience* comes from working with prospective teachers in ways that do more than simply (re)confirm their existing beliefs about teaching and learning. This means learning to support and challenge prospective teachers so that they may be willing to suspend their beliefs in order to entertain alternative approaches to pedagogy.

This tension played out in two main ways in the Biology methods class. One way was through the deliberate pedagogical structures that I employed to support prospective teachers' knowledge building through experience, while the other relates to an unplanned activity of sharing personal experiences of teaching with one prospective teacher, Lisa.

BUILDING ON EXPERIENCE THROUGH DELIBERATE PEDAGOGICAL STRUCTURES

I introduced a variety of tools into the Biology methods classes to help students begin to focus on the nature of their experiences and to draw meaning about teaching and learning from these experiences that they may not have previously considered. These tools included: a Personal Learning Review (PLR) completed by students early in the year to elicit their experiences and expectations of learning Biology; the Open Journal that I maintained on the worldwide web (that students were invited to respond to; and a Drawing Task in which students were asked to draw and interpret a picture of themselves teaching a 'typical' Biology class from their field experiences. I also devoted a considerable portion of the Biology methods curriculum to peer teaching in order that students could have experiences of teaching each other and then 'unpack' these shared experiences through examining their perspectives as both learners and teachers.

My decision to include these tools was based on my belief that students needed to know about themselves as learners in order to understand more about how they were likely to operate as teachers. As Brookfield (1995, p. 49) points out, ". . . our manner of teaching is, to a great extent a direct response to how we were taught . . . We attempt to replicate the things our own teachers did that affirmed or inspired us as learners".

As their teacher educator, I wanted to know about these students' past experiences as learners so that I might be more sensitive to their perspectives, and use this knowledge to inform curriculum experiences that might expose students to new approaches to learning Biology. My view of learning Biology as a process of conceptual change, also meant I recognised that prospective teachers' prior views would not easily shift, and that examining, discussing and challenging their views were important means of developing new kinds of knowledge and new ways of understanding about pedagogy. What follows then, is a description of each of these tools, their use in Biology methods classes to help bring out, and build on, prospective teachers' experiences and some of the issues that I encountered along the way as I sought to value and reconstruct with them, their experiences of learning to teach.

PERSONAL LEARNING REVIEW: LOOKING BACK ON EXPERIENCE

The Personal Learning Review (PLR) was a tool that served the purposes of both teaching and research. The PLR task required Biology methods students to answer a series of questions about their past experiences as Biology learners at school and university. The teaching purpose of the task was to help build a picture of prospective teachers' entering assumptions about Science, teaching and learning. I wanted the prospective teachers in the Biology methods class to begin to recapture and explore the experiences and influences that shaped them as learners. I anticipated that through the process of revisiting and articulating their memories that they might begin to know more about what influenced and motivated them, and perhaps, how they might be likely to act as teachers. In this way I was valuing their experience, letting them know that the experiences they bring matter. However, eliciting memories and influences is a double-edged sword, since surfacing these memories may simply lead to reconfirming what worked for them, which might then reinforce the same in their own teaching. Bullough & Gitlin (2001) note the difficulties associated with such forms of life writing, ". . . because much of what life writing reveals is self-confirming, strongly valued as part of self and impervious to change" (p. 24). This situation underscores the tension described in this chapter, between valuing students' experiences and challenging them to question and look beyond these experiences.

Students' responses to the PLR task generally revealed a group of conscientious learners who had mostly experienced a transmission style of Biology teaching at school and university, with practical activities as an enjoyable exception to the didactic routines of traditional Science learning.

Students generally were motivated to continue their Biology study as a result of their personal connection with an aspect of the subject (e.g., studying the environment, how bodies work, disease) or a teacher's friendly encouragement. My intention was for students to revisit the PLR task several times throughout the year as a stimulus to begin to recognise differences between themselves as students (what motivated them, what led to their success in the subject) and the students they were now teaching.

I anticipated through this process that prospective teachers might start to see that what was helpful for them as learners, may not actually be what is most helpful for their students' learning and, as teachers, they needed to respond to the varied needs of their learners not just reproduce their own successful experiences (i.e., to reconstruct experience). While this was my intention, a different story emerged in practice.

Biology methods students revisited the PLR task twice more during the year (at the end of semester 1 and at the end of the academic year). For the most part, they could readily identify differences between themselves and their students and the influences of their past experiences on their teaching. However, knowing about themselves did not lead, as I had anticipated, to them seeing an immediate need for rethinking, or 'reframing', their practice. Instead, these activities seemed to help them affirm a view of themselves. It seemed as though prospective teachers now had a way of rationalising their practice, (consistent with the "self-confirming" idea of Bullough & Gitlin stated previously). Practicum experiences further served to reinforce prospective teachers' views of themselves, influenced by supervising teachers and school students' expectations of how these new Biology teachers should behave in their role. Informal conversations with these prospective teachers during practica led me to recognise the difficulties associated with their questioning (let alone shifting away from) long held beliefs. Under pressure, long held patterns of behaviour surface (Mason, 2002) and for these new teachers the need to perform expected actions and their memories of success served as powerful influences on their thinking.

Drawing Task – A Way of Seeing into Experience

I introduced the Drawing Task following the first teaching round. I asked students to draw a picture of themselves teaching a typical Biology (or Science) class from the practicum and to include some of the things they typically said, and thought about, while they were teaching. The purpose of this task was to provide an opportunity for students to 'unpack' their practicum experiences using a novel approach that I hoped might highlight particular issues, concerns or insights from their teaching and so help them begin to articulate their developing knowledge about teaching.

I knew that some form of debriefing activity was a common post practicum task across all subjects in the teacher education program and I suspected that as a consequence, debriefing could simply be seen as a routine that carried reduced meaning for students the more it was encountered; even though they generally enjoyed sharing stories of practice with each other. The challenge for me was to engage students in some form of meaningful analysis of their experience, at the same time acknowledging and valuing their experiences. So, I decided to introduce a more innovative approach to debriefing through drawing.

I anticipated that the unusual nature of the task (drawing) might open up some otherwise subconscious assumptions about teaching and help students probe the nature of their experiences more deeply – similar to Richards' (1998) work with prospective teachers' self-portraits. At the same time I was modeling a teaching procedure ("draw your thinking about a topic", White & Gunstone, 1992) that I hoped these students might consider using in their classes.

Students completed the drawing task individually then shared their drawings in small groups, with the instruction to identify issues, look for similarities and differences amongst the group's drawings and to ask questions for clarification of one another about the drawings. Finally, I asked each student to write a short reflective summary underneath their picture that captured any insights about themselves and their teaching that they had gained through the activity, and/or general statements about teaching arising from the shared experiences of the group.

It was not difficult for students to find shared concerns from their pictures: dealing with classroom management issues; fear of not knowing the required content; apathy from school students; and, lack of support from supervisors, all readily emerged. In this way, one purpose of the task was realised – the group was able to identify common concerns. In fact, it was difficult to shift some individuals beyond sharing stories or offering each other advice about dealing with various issues, (despite regular encouragement from me to go a little further and say what they had learnt about teaching Biology from their discussions). Others though, were able to draw insights and construct general statements about their teaching from what they drew and discussed. One group progressed to the point of developing several generalisations:

- Structure in lessons is highly valuable.
- Believing in students is effective.
- The teacher cannot assume in advance the students' level of knowledge.
- Requesting that students respect each other is helpful for effective lessons.

As a collection of statements about teaching these seem perhaps simplistic and obvious. However, what matters is that the students articulated these ideas from their own experiences rather than me, the teacher educator, supplying my interpretations of their experiences, (or telling them about my experiences of teaching). These prospective teachers were able to identify collective concerns then build on them in a manner that extended beyond their individual contexts and experiences. This is similar to that which Loughran (2002) noted when he highlighted the value of student ownership of knowledge production in learning about teaching:

> . . . if the focus is genuinely on the student teacher as learner, then it is their ability to analyze and make meaning from experience that matters most – as opposed to when the teacher educator filters, develops, and shares the knowledge with the student teachers . . . the knowledge developed may well be the same, but the process in developing the knowledge is very different. Who is doing the learning really matters . . . (p. 38)

Student feedback about the drawing task indicated that they had mostly understood its purpose as a means of reflecting on the issues encountered individually and as a group, although they were mixed in their enthusiasm towards it. Pleasingly, one student recognised the act of drawing as a way of bringing to a level of conscious awareness ideas about oneself as a teacher that might otherwise pass unnoticed, (*"When you have to draw something . . . the things that come out may not be the things that you thought about before"*). After discussing her drawing with others in her group this student came to realise that the mood of her Science class and their inability to settle was probably due more to their poor attitude towards Science, rather than a personal attack on her.

In contrast, another student reported that: *"It [drawing task] didn't bring out anything that I didn't already realize . . . [and] I still don't have any solution to the issue"*, but then later admitted, *". . . maybe I did learn something, that I have to stop being so defensive about constructive criticism"*. I found this student's response surprising because up until this point she had seemed particularly steadfast in her beliefs and resistant to possibilities for change. Perhaps this was evidence that she was beginning to shift a little. Responses such as these helped me realise that for at least some students, the task had moved them beyond their initial thinking about their experiences to construct new meaning that had real applications for their practice.

PEER TEACHING: RAISING AND CHALLENGING INTERPRETATIONS OF EXPERIENCE

While the PLR and Drawing Tasks were primarily designed as ways of raising prospective teachers' self-awareness of the way they functioned as teachers, the peer teaching was intended not only to raise, but to also challenge the way in which students thought and acted as Biology teachers and learners, and to consider new possibilities for their practice that were consistent with meaningful learning.

A continuing theme in my working with students in this way was to set up experiences from which *they* examined *their* shared experiences as a means of building their understanding of teaching. I saw my role as one of encouragement and support, and also as 'provocateur', as I recognised opportunities within their teaching to challenge them to interpret situations in new ways. However, I soon found that these two roles did not fit together neatly. It was difficult for me to effectively support and challenge students.

The disparity that emerged through this situation was that if I supported the approaches they chose then it might reinforce their belief that what they were doing was appropriate and effective. If I challenged them too much perhaps they would reject what I had to offer because they felt threatened or vulnerable, and as a consequence remain with what was comfortable for them. This then is the essence of the tension as I lived it with my students. And, although I espoused an approach of challenging students' ideas but not telling them what to think, I found that in practice, I was actually very keen for students to see what I saw in their experiences; even when that meant encouraging them to see their teaching as problematic when they felt as though it had been good or successful. The following vignette, constructed from a Biology methods peer teaching experience is designed to illustrate this tension as I experienced it.

Vignette: Why can't she see what I want her to see?

Gemma was teaching the group about genetics. For the most part, her session consisted of telling us a series of definitions of genetics terms. Her manner was lively, but it did not move beyond a transmission approach. As I sat and watched I grew increasingly concerned.

"C'mon Gemma, give it up!" I thought. "Maybe we know some of this already? You need to find out what sense we are making of these terms."

The students listened politely. I was unsure about whether to intervene with a question or wait until she had finished. Knowing Gemma though, my intervention may well have led to longer and more fine grained telling, so I decided to wait. When the teaching session concluded I asked her, "What could you pick up about our learning?" She responded confidently, "I looked around and I could see that everyone got it pretty well."

Initially, her response stunned me. How could she possibly think that? Did she really believe that she could accurately monitor learning in this way? Perhaps though, she felt put on-the-spot by my question and her manner of responding was a way of saving face.

I wanted to persist with my questioning because I wanted her to see that there was a problem here, that she had made some assumptions about our learning. Yet, at the same time, I did not want to publicly embarrass her. I paused and looked around the room. Maybe one or two of the students might be prepared to pick this up for discussion? As though on cue, Dan muttered quietly, "I didn't." I capitalised on the opening he offered.

"Sorry Dan, what was that?"

"Yeah, I didn't get it. I knew I was supposed to know what these definitions meant, but I didn't. It's been a long time since I've studied this, and I've forgotten most of it."

Gemma turned to me with a cynical look. Her expression seemed to suggest that Dan was simply playing devil's advocate on my behalf. I tried to open up the conversation and encourage Dan to talk with Gemma so that she might start to consider more about the perspectives and needs of the learner.

"Maybe there's something you might like to ask Gemma now? To explore a little more?" I suggested to Dan.

"Umm, I'm not sure about base pairing and how that is different in RNA and DNA."

"It's on the note sheet that I handed out. See, on page four", replied Gemma.

"Yeah, but I don't get it", he intoned.

"I think that it is difficult to explain everything in one short session. This is perhaps something you need to go and look up now for yourself", she responded firmly, closing down any further conversation between them.

Later, in discussing this episode with a colleague, I puzzled over the issue of my need for Gemma to recognise a difficulty with her approach and her apparent unwillingness to see what I wanted her to see. I was keen for her to know that it is not a simple relationship between telling people things and them learning or understanding it (highlighting the related tension between *telling and growth*). Through our discussion, I began to realise my assumption that simply raising this issue with Gemma did not mean that that she automatically recognised it as a problem or that she would be prepared to explore what I saw in her teaching approach. She needed to experience this as a problem herself, to feel it as real and problematic. Perhaps then it would make more sense and possibly, lead her to some new understandings of practice.

The central issue of the vignette, and of the tension described in this chapter, is my struggle between wanting students to make their own meaning from their experiences, and my concern to steer them in the direction of particular kinds of understanding about experience that I believed to be important. Because Gemma did not recognise or value my meaning, I tried to exert a pressure on her to come around to my point of view. But as Korthagen (1999, p. 101) observes, "The best way to stop change in a person is to try to impose it on him or her". Korthagen proposes that it is this discrepancy between the perspectives of the teacher educator and the student that inhibits the student's growth. Similarly, Schön (1987) identifies the importance of teacher and student working towards "achieving a convergence of meaning", and making opportunities available for the prospective teacher to feel what it is like to be in a particular situation, and in feeling, and reflecting on these feelings, generating "meanings . . . not previously suspected" (Schön, 1987, p. 117). Hence, for teacher educators to have an influence in developing prospective teachers' knowledge through experience, it is first important to help students recognise that dilemmas exist, so that they may feel more confident to explore implications for themselves and in the process, reconstruct their knowledge about teaching in more personally meaningful ways.

A PEDAGOGICAL SOUNDING BOARD

A different way in which the tension between valuing and reconstructing experience played out in the Biology methods class was through the e-mail conversation shared between Lisa and me. Our e-mail relationship involved a great deal of scrutiny of how each of us behaved as teachers, implications for our students' learning and the meanings we attributed to certain events. Although unplanned and with no clear purpose beyond exploring ideas and experiences together, this arrangement offered a useful pedagogical sounding board for each of us, and became a powerful tool for understanding practice. Exploring practice together in this more personally responsive way was very different from the approach I employed through teaching procedures and structured tasks with the Biology methods class.

A strong sense of openness and trust characterised exchanges between Lisa and me, as well as the enjoyment of finding a like-minded person who shared the pleasure of talking about pedagogy. Interestingly, Lisa rarely asked me direct questions about teaching nor for my opinion about issues she was grappling with, although I am sure that she was keen to receive my comments. From time to time I worried that I was not giving her the kind of (written) support that she may have wished for, however, as the year progressed I came to understand that Lisa did not need me to sort through her experiences and provide answers or recommendations about her thinking in the way that I anticipated she might; instead she needed to know that I was listening, and that I cared to know what she thought. Her trust in my support meant that she was free to sort through her ideas by herself.

Initially our e-mails grew out of Lisa's wish to keep a record of her experiences and my interest in learning about how she was experiencing her teacher preparation. However, our exchange evolved to serve several equally important, additional purposes. These included acting as a support and encouragement for each other; identifying, exploring and revisiting issues about teaching and seeking deeper meaning from experience. (Interestingly, these functions correspond closely to those described by Russell and Bullock (1999). Russell, a Canadian teacher educator, and Bullock, a student from his Physics Method class engaged in a shared exploration of their experiences via e-mail exchange). This illustrated how powerful perspectives on learning and teaching can develop through revisiting and re-evaluating experience through another's perspective. The following examples explore these purposes more fully.

Support and Encouragement for Lisa

Several times during the year, Lisa told me (in person or in writing) how helpful it was for her to be able to write about her experiences and to share them with me, knowing that I cared to read what she had to say, and that I took her ideas and experiences seriously. I was conscious of the considerable risks that she was taking in her teaching and wanted to affirm and encourage her, to help her feel that what she was doing was worthwhile.

Date: Sun, April 29 2001
From: Lisa
To: amanda.berry@education.monash.edu.au

Can't tell you how useful this is for me, seeing my ideas up on the screen, having time to think them through, particularly after an awful session like today. I feel really privileged to be able to talk about all this stuff at any time – it's such a help.

Date: Mon, April 30 2001
From: Mandi
To: Lisa@student.monash.edu

. . . I think you are really brave because you are trying out some incredibly challenging stuff that experienced teachers don't even recognise is a problem in their practice . . . Asking yourself about what is the role of the teacher in the power relationship in the classroom/ How does that relate to science teaching and learning/ How do you interpret the curriculum in a useful way for students etc. is really hard stuff. Don't be too hard on yourself, there is so much going on . . .

Our e-mails allowed us to sort through new ways of thinking and affirm new ways of acting as teachers and motivated us to persist through difficulties.

IDENTIFYING, EXPLORING AND REVISITING ISSUES

Lisa spent a great deal of time puzzling over issues that grew out of experiences she had at university or in school. Often she would include a title to her e-mail that identified the issue that she was exploring in the ensuing note. Many times her writing was a working through of her thinking, although not necessarily working through to resolution. An example of this was her puzzling over the issue of intrinsic and extrinsic motivation in learning.

Lisa linked a discussion from Biology methods, in which students were discussing their use of lollies (candies) as incentives for improving their students' motivation for learning, with her experiences of a Science workshop in which the lecturer stimulated Lisa's motivation for learning through the teaching procedure itself, rather than using an external 'bribe'. She knew intuitively that intrinsic motivation must be a better form of learning, although at that point, she was not yet able to clearly articulate the reasons underpinning her thinking.

Subject: HOW CAN LEARNING BE THE REWARD FOR LEARNING?
Date: Sat, March 3 2001
From: Lisa
To: amanda.berry@education.monash.edu.au

. . . Megan brought this up in Biol today when she was talking about how she sought to control her class from WPC[1]. Her group bribed the kids with lollies. I am pretty judgmental about this whole concept. I really don't like it at all. Sugar is so bad for learning that I can't see how it can be good to hand it out as a reward or a bribe. It makes the kids hyper when the sugar kicks in, then makes them lethargic when the sugar low hits. I am just so opposed to this idea. Someone else introduced the idea of using other things as bribes – activities etc. I think learning can be an incentive. What I mean is, an idea is introduced, and will be discussed or revealed later – that is a kind of a bribe. This is what Dick did[2] – he introduced the idea, got us engaged, and bribed us to come back next week by offering to tell us the answers next week. He initiated the intrigue and used the knowledge as a bribe. I would much prefer to try to do things like this in the classroom than use lollies. What else could I use? I just don't really like the idea of physical bribes and rewards – things like food, free time, play time etc. I just think it is wrong, but I can't really explain why.

---

[1] Local school.

[2] Early in the year, a predict-observe-explain teaching procedure (dropping two balls of different masses from the same height) is modelled for the student teachers in a science workshop. Students are asked to predict which ball will hit the ground first and why, before the demonstration is run. The lecturer ran the workshop over two sessions, delaying his explanation of the outcome of the demonstration until the second week, and thereby creating, as Lisa noted, some motivation for learning (and returning to class!) the following week.

Lisa raised and revisited issues that caught her attention throughout the year. Mostly, these issues focused on student participation, motivation and choice in learning. E-mail writing served as one of several different ways that Lisa used to make meaning from her experiences.

Seeking Deeper Meaning

Lisa's explorations of her various experiences led, on occasion, to insights that she shared with me. Sometimes these insights came as an epiphany, while at other times they formed slowly, crystallizing over several exchanges. Through her reflections, Lisa came to recognise and articulate the frames of reference she was using to think about teaching and learning and, as a consequence, she began to reconsider and restructure, or "reframe" (Schön, 1987), her views. For example, in her e-mail of August 7 (below), Lisa realised that her existing 'frame' for thinking about teaching and learning was more narrow than she had imagined. Then, on August 12, she described a shift in her thinking as she reframed her perceptions of teacher as leader.

Date: Tues, Aug 7 2001
From: Lisa
To: amanda.berry@education.monash.edu.au

. . . You know, I think I have just had one of those moments, when you see things. I have spent most of this year thinking about teaching and learning in a very rigid and uncompromising way and I have only just realized that this second.

Date: Sun, Aug 12 2001
From: Lisa
To: amanda.berry@education.monash.edu.au

. . . Am slowly starting to realise why I am the teacher! . . .This might be really obvious to most people, but it is going to take a long time to sink in that I 'know best' to some degree . . . It hadn't dawned on me until that moment that being a teacher meant being a leader, and I'm not really comfortable being a leader . . . But if I want to be a good teacher, I must be a good leader.

These e-mails provide evidence that Lisa was playing an active part in her own learning, reflecting on and interpreting her experiences as they became important for her. This process, Barnes (1998) argues, is critical for prospective teachers in shaping their classroom practice, and is far more likely to influence how prospective teachers will act in the classroom compared with the well meaning, but mistaken practice of teachers (at all levels), of trying to impose their ways of thinking on students as means of effecting their change, or telling them what they should have seen or learnt in a given situation.

For beginning teachers . . . [it] is no use to offer them the thoughts of an experienced teacher, for these will merely be reinterpreted in the light of their preconceptions. *Reframing* by the student teachers themselves is crucial. (Barnes, 1998., p. xiii) (author's italics)

In the light of Barnes' comments, I find it particularly interesting how little I (explicitly) imposed on Lisa my ideas and beliefs in my correspondence with her, or tried to interpret what she was thinking or feeling. Much of the time, I was simply by her side as she examined her experiences. My way of being with Lisa contrasted strongly with my approach to teaching Biology methods classes where I frequently felt the need to impose my meanings and push the direction of the learning toward what I wanted, rather than trusting students to make their own meanings as Lisa so capably did. During our second interview Lisa told me it was 'simply' listening to her and asking her "why" that was a most effective stimulus to her reconsideration of her thinking and practice.

> Lisa: . . . mainly the biggest thing that I think has helped me learn this year is having e-mails between us . . . it is just so amazingly helpful and mainly because I am just writing down my own ideas and going blah blah blah blah and you go "oh why?" and that is the push . . . So you're not saying "Oh yes, I agree and you know this is what I do . . . ". You're just helping me go through those thoughts and think about things a bit more . . .
> (Lisa Interview 2: 251–253)

Lisa's words underscore the trusting nature of our relationship. As part of our trusting relationship, I was also patient with Lisa, in a way that illustrates another facet of the tension between valuing and reconstructing experience in practice. Too often in the Biology methods classes I expected that my intentions for students' learning would be realised within the space of the method session and that I would be able to see concrete evidence of their changed thinking on the spot. In other words, I expected them to build (or rebuild) knowledge from experience as I watched. In hindsight, this realisation surprises me; since my espoused belief about learning to teach is that it is a process that evolves slowly over time. In practice, however I recognise that I behaved in ways that were inconsistent with this belief (thus highlighting the linked tension of *action and intent*).

INSIGHTS INTO MY TEACHING THROUGH LISA'S EXPERIENCES

At the same time that Lisa's e-mails provided a context for exploring her thinking about her teaching, she was offering me valuable insights into mine. Her accounts of the Biology methods classes, including her observations of my behaviours helped me better understand the experience of Biology methods from a learner's perspective and in the process, build my knowledge of practice. Sometimes her feedback surprised me. For example, on one occasion she noticed that student contributions to class discussions seemed to always come from the same (few) people: (e.g., *"I see this happening in Biol a lot – it's always the same people who make comment. What are the quieter people thinking?"*), on another that I was not practising what I was preaching (e.g., *"You know that thing you said about how when a few people answer, the teacher makes an assumption about what the whole class knows? You do that.")*. At other times, her feedback affirmed and encouraged me to know that at least one person was catching the messages that I intended the class as a whole to hear.

This feedback and support was particularly helpful on those occasions when I felt lost and uncertain about the worth of the teaching approach I was trying to implement. Lisa also helped me understand more about the perspectives of other students as she compared her own ways of thinking and acting with others who were different. Just as Russell (1999, p. 143) commented that his relationship with Bullock: ". . . helped me understand particular features of my teaching more fully . . . and take points further than I might have on my own", so too, I derived similar insight through Lisa's relationship with me.

One striking example of the way in which Lisa helped me build my knowledge of my practice was through her observations of my expectations for students' learning in Biology methods classes. In particular, my expectation that students would understand, appreciate and accept my agenda for their learning about teaching. This is significant because it illustrates how difficult it was for me to act in ways that were congruent with my espoused beliefs about teaching teachers. Even though the tools and activities I prepared for Biology methods (described previously in this chapter) were designed to surface, challenge and extend prospective teachers' beliefs about teaching, and to help them make sense of their own experiences, at a deeper subconscious level, I still wanted students to see what I saw and to take on my purposes as their own. Two excerpts from Lisa's e-mails illustrate this point.

Date: Mon, April 30 2001
From: Lisa
To: amanda.berry@education.monash.edu.au

. . . It would be great if we are able to remember what we are doing and why for all the time we are teaching, but even if we only think about this some of the time, I reckon it will be a great thing, and you will have helped us to do it.

And then a few months later, another reminder:

Date: Thurs, Aug 30 2001
From: Lisa
To: amanda.berry@education.monash.edu.au

. . . Perhaps we [students] take from an experience what we need at the time.

The tension played out for me as Lisa helped me recognise a problematic situation within my practice. Yet, I was unable to move beyond this recognition, to act differently. Just as prospective teachers needed time to re-examine and restructure their understanding of experience, so too did I.

Earlier in this chapter, I noted the importance of the first step of recognising that dilemmas exist in practice. 'Living' the tension as I did, gives me greater insight into the challenge inherent in its meaning. This tension, just as the language which conveys its meaning suggests, had a great deal to do with me experiencing in my own practice, the very things I intended to be learning opportunities for students.

SUMMARY: WHAT DID I LEARN FROM EXAMINING THIS TENSION WITHIN MY PRACTICE?

Few teacher educators would deny the power of experience in prospective teachers' learning to teach. However, opportunities to learn from experience can be unintentionally hampered by those in positions of authority (school based supervisors, university teacher educators) (Britzman, 1991) who impose their well rehearsed interpretations of experience on situations rather than supporting and challenging student teachers to find meaning for themselves. "I know because I have been there and you should listen" (Munby & Russell, 1994, p. 93) is the message regularly conveyed either explicitly or implicitly to prospective teachers about the nature of learning to teach. This means that the impetus for prospective teachers to take an active part in their own development is taken away; even though it may occur inadvertently.

Teacher educators need to develop a sensitive appreciation of the individual needs of prospective teachers so that they can support the process of meaning making:

. . . providing support for the personal courage required to question the completeness of longstanding explanations for personal experiences, and considering alternative explanations for those events. (Korthagen, 2001, p. 365).

Through an examination of my experiences I have identified different aspects of the tension of valuing and reconstructing experience. Being aware of the tension made me sensitive to how I worked with the prospective Biology teachers, and more aware of how the tension played out in the practice of others – although it did not mean that I was able to recognise all of the ways in which this tension was operating within my practice at the time. Often I believed I was successfully working towards this goal without realising that my actions crafted a different picture. Even when I did recognise the tension in practice, this did not make change easy or straightforward. As Senese (2002) noted, so too I found that acting differently to that which I intuitively felt was needed, was a difficult task; despite believing I knew that all along. Korthagen (2001) similarly pointed out this issue:

What seems obvious to the teacher educator is not so to the student teacher. What to us seems directly applicable in practice appears to be too abstract, too theoretical, and too far off for someone else. What seems to us evident and easy to understand does not get through to the student. No matter how carefully we consider the problem we do not find a way into it. Or we may seem to have found such an entrance and gotten through to the student, but afterward it becomes clear that the student has not even tried to carry out the resolution. Apparently there is an unbridgeable gap between our words and the students' experiences. (p. 22)

This is indeed the essence of this tension in action in my teaching about teaching.

Chapter Eleven

REVISITING AND SUMMARISING THE TENSIONS

INTRODUCTION

The previous chapters of this section have isolated and examined a set of tensions, and my learning about them, from within a one year self-study of my practice as a Biology teacher educator. Presenting the tensions in such a way may unwittingly foster a view of each as distinguishable and separate and, although necessary for the purposes of explanation for others, is certainly not how I experienced them in the real world of practice. In this final chapter of this section, I want to distil my learning from the tensions in a more holistic manner. In so doing, I hope to illustrate how my professional knowledge developed as a consequence of the recognition of, and interaction with, these tensions in my teaching about teaching. I therefore revisit and reconsider the notion of tensions in light of their interconnectedness and explore some of the implications inherent in the development of different forms of knowledge. I begin by briefly reviewing the notion of 'tension' and its relevance to the nature of this self-study.

REVISITING THE NOTION OF TENSIONS

The notion of tensions is intended as a way of representing and better understanding the elements of ambivalence and contradiction so intrinsic to the complex nature of pedagogy. In teacher education in particular, tensions are helpful as they are borne of attempts to match goals for prospective teachers' learning with the needs and concerns that prospective teachers express for their own learning. Teacher educators are confronted by a role that: "... is experienced quite deeply and frequently as a series of dilemmas" (Stronach, Corbin, McNamara, Stark & Warne, 2002, p. 13). Conceptualising practice as a set of tensions supports the diverse, complex and uncertain nature of practice, as it takes into account the ways in which teacher educators (including me) talk about their practice and enables practice: "to be understood in its complexity, plurality and inconsistency" (ibid., p. 13).

BRINGING THE TENSIONS TOGETHER

While represented separately for the purpose of describing them in this book, each tension is interconnected in the real world of practice, and between the tensions themselves there are inevitable links. These tensions of practice are therefore intertwined in such a way that understanding which is operating at any particular time is difficult; since each tension impacts on (an)other(s). Different tensions become highlighted as different aspects of teaching about teaching assume more or less importance and as the pedagogical purposes shift within and across experiences and episodes. In essence, there is ebb and flow between the tensions such that they may all well exist at once, but rise to the surface in different ways at different times depending on the situation and the way that it may be "played out".

For example, *telling and growth* is concerned with balancing the delivery of information about teaching with providing conditions for growth. Such conditions for growth include a trusting environment in which prospective teachers are willing to take a risk. This then highlights the affective domain associated with telling and growth – that of *confidence and uncertainty* – whether prospective teachers (and teacher educators) have the confidence to push ahead (or not) with new ways of working. The development of confidence leads to one's ability to move away from safe, known ways of operating and challenging oneself to behave in new ways, hence the tension of *safety and challenge* is invoked. Throughout this process one must be sensitive to the individual needs and concerns of prospective teachers so that how one works as a teacher educator is responsive to situations as they arise (*planning and being responsive*), *valuing and reconstructing* the particular *experiences* of the individuals involved. Lying at the heart of all of this is the congruency between *action and intent*, what the teacher educator and prospective teachers want (or need) to happen and how they actually behave in order to achieve their goals (which may or may not be consistent with their beliefs about what they want to achieve).

The interconnections between tensions are illustrated through the following vignette. The vignette is designed to show how interconnections between tensions become apparent through the practice context, and how situations emerge in teaching in which the tensions may become clear and explicit, or be embedded in possibilities inherent in decisions about what to do (or not to do) in teaching itself. (Tensions appear in bold type within the text.)

Vignette: Interconnecting tensions in practice

"I don't like group work."

    "Me either. I'd rather work alone."

    "I don't mind it because everyone else does the work for you."

    "Well, I like it. It gives me a chance to learn through discussion and sure beats listening to a lecture!"

These words just fell out of the student teachers' mouths as I organized them into groups.

I had set up a 'jig-saw' activity using the topic of disease transmission as a way of raising issues about starting a new unit of work. The activity involved several different 'stations' each with a separate task, and all related to different aspects of the transmission of pathogens.

[The structure of the jig-saw method is that each member of the 'home' group accepts responsibility for a particular task (the number of students in a group being equal to the number of tasks). Each of the people from the different home groups who have the same task get together to form an 'expert' group and together they work out their response to that task. When they have completed their response to their task, all of the members of the expert groups rejoin their home groups and teach their peers what they have learnt. Therefore, each group member becomes an expert in one aspect of the topic that they teach to their home group. Together, the home group develops an understanding of the total topic as they build their knowledge through the combined experience of being both teachers and learners (**telling and growth**).]

As the jig-saw came to a close, I began to focus students' attention on the teaching and learning they had experienced.

"So what features of the activity helped, or didn't help, you to learn?" I asked. That opened up a torrent of responses. It was obvious that students had enjoyed the activity and readily distinguished between their approaches to learning in the expert group and their home group.

"I didn't do much in the expert group because I thought I'd just get the answers and go back to my home group. But when I went back I couldn't really explain what I was supposed to know about so I felt bad that I had not taken it seriously!" Kellie blurted. "Thanks for that. Anyone else?" I inquired (**safety and challenge**).

"After working out the ideas in the expert group, it was interesting to be asked questions about my ideas back in the home group. I think explaining something to someone helps you understand it better," Joanne remarked.

"Yep, me too," added Nick. "In fact questioning in the group was different from the type of questions you get in a class. Like I told the group I didn't understand some of this stuff. I'd never say that out loud in front of a whole class though! . . . Huh, I've just said it. How 'bout that!" (**confidence and uncertainty**).

The discussion continued for some time and the ideas raised highlighted a range of features of learning that, based on their experience of doing the jig-saw, more than covered the things I would have wanted to raise (**telling and growth; planning and being responsive**).

However, in this case, the ideas carried personal meaning for the students because they were involved in the learning (**valuing and reconstructing experience**).

Then I asked about the way I had taught the jig-saw.

"You didn't teach. All you did was tell us what to do and then you walked around and watched us do all the work," Lauren asserted.

"And you certainly didn't interfere with us. You didn't tell us what was right or wrong, you left us to work it out in both our groups," said Sue (**action and intent; confidence and uncertainty**).

"No, I think you're missing the point," Lisa interrupted. "The teaching was in preparing the different tasks. If you think about each thing, there was the chunk of information that one group worked on, like a comprehension task. Then there was the graph thing where you had to work out about the time it took for an infection to take hold. Then there was that big picture thing, not sure how you'd describe that, I really liked the way we had to try and work out what the lines of defense were just from the pictures."

"Yeah, I didn't know that mucus had so many functions!" said Jeff (**action and intent**).

With that, a deeper dissection of the 'teaching' ensued and the students started asking me why I did things the way I did. "Why didn't you come around and check on whether we were doing things properly?" "How come there wasn't a teacher summary at the end?" "What if we didn't get the right information?" were some of the questions that were thrown at me as the students began to carefully examine the way the teaching had been organized and conducted (**safety and challenge**).

As the session finished and students were leaving, I felt as though they had become aware of both their learning about learning and their learning about teaching. It was a nice feeling to have a session work the way it was planned (**action and intent**).

This vignette illustrates one approach that I take to learning about teaching. It places at its centre the chance for participants to experience their learning through *doing* a task so that they might better understand the nature of the task and hopefully, consider how their students might experience it (i.e., creating conditions for growth rather than simply telling prospective teachers about the jig-saw teaching procedure and the Biology content).

As a consequence of prospective teachers having this experience as learners they are better able to comment on what it is like to do this kind of group work from their different perspectives and that different individuals differ in their responses to the task. As their teacher educator, I choose to step out and take the risk of teaching and debriefing in this manner because I believe that such an experience provides a basis for the development of meaningful learning about teaching. I also trust that I can work in this way. I encourage my students to step out and take a risk in talking together about my teaching and their learning, and in this way I hope that they will come to acknowledge and build on the experiences they have as learners.

Bringing the tensions together in this way also illustrates the rich variety of implicit, interconnected aspects of practice (e.g., philosophies, attitudes, skills, concerns, teaching procedures and so on) that support and interlink the tensions. In teaching about teaching it is not always possible or practical to construct sessions with such explicit links to tensions because, in reality, the tensions are derived *of* practice, not necessarily pre-structured *for* practice. Thus the process of developing my pedagogy of teacher education lies more in articulating my learning through analysis of practice for knowledge production, rather than as theory that directs practice. In this case, practice informs theory.

While each of the abovementioned aspects of practice is important in teaching about teaching and has become articulable in defining my pedagogy of teacher education, the development of particular teacher educator attitudes in creating specific kinds of learning experiences with prospective teachers has also become apparent to me as a consequence of my learning through this self-study. This then illustrates how a self-study of this kind continually highlights new areas for research as my teaching and research interact in a dynamic and holistic fashion.

Teacher Educator Attitudes

As a result of analysing my teaching about teaching through the frame of tensions, I have come to recognize that across the tensions, particular teacher educator attitudes are highlighted and recur. These attitudes include a commitment to:

- caring (such as that described by Mayeroff (1971), "To care for another person, in the most significant sense, is to help him [sic] grow and actualize himself") (p. 1);
- paying attention to the *individual* needs of others – which is different from the needs of a particular group. It means, "responding sensitively to the specific needs, hurts and potentials of specific students" (Mayes, 2001, p. 489);
- genuineness and honesty;
- taking risks and exposing one's own vulnerability. One risk is in risking relationships with students. (When one defines oneself as a teacher through relationships then this risk can be very demanding e.g., see Schulte, 2001); and,
- trusting in oneself and one's students.

I briefly explain the influence of each of these attitudes in my practice, as follows.

Caring

When a caring relationship is established opportunities for personal and professional growth are enhanced. Caring means being attentive and receptive to others' needs and concerns and refraining from immediately imposing one's own agenda on the situation (Noddings, 2001). In my interactions with students in the Biology methods class, I found it difficult to establish caring relations when I did not look beyond my own agenda to try to understand more about the expectations and motivations of individual students. For example, I struggled to enter into a caring relationship with Bill because of my feelings of disapproval related to his approach to teaching, which in turn may have reduced his opportunities for growth. In contrast, my relationship with Lisa was attentive and receptive to what she was experiencing, and Lisa recognised this, which supported her growth.

Noddings (2001) proposed a view of caring that is not as a person who possesses certain virtues, but one who "more or less regularly" establishes caring relations. This view acknowledges that relations of care may fail in one situation and succeed in another. Caring happens over time, for more than one person simultaneously, and caring can conflict with other demands and expectations.

Paying attention to the individual needs of others

Being sensitive to the range of needs (and concerns) of others means listening to prospective teachers, accepting their resistance to certain things and coming to know more about how it feels to be that person (Rogers, 1969). An important prerequisite to connecting with and responding to the needs of others is a sense of self-understanding. When I was busily preoccupied with my own personal agenda (for instance, when I was caught up in feelings of confusion and unsure about how best to proceed in a particular situation) it was difficult for me to see beyond myself to attend to what different students were thinking or feeling. Even when I believed that I was genuinely paying attention to the individual nature of their concerns, I often viewed the individuals in my classes as, "younger versions of [my]self" (Trumbull, 2004, p. 1221) and consequently treated them as I would have wanted (or expected) to be treated or, projected my self-concerns onto the students so that I may have misinterpreted how particular individuals were thinking or feeling.

When the learner is not like the teacher, this misplaced empathy may be unproductive in terms of prospective teachers' pedagogical growth. I experienced difficulties connecting with, and acting in a responsive manner to, some of my prospective Biology teachers, thus not necessarily supporting their learning. Putting aside my own thoughts, feelings and assumptions to view a situation from another's perspective is a challenging task and one that relies on first recognising and accepting aspects of oneself. I struggled constantly with this aspect.

Genuineness and honesty

Korthagen noted that, "The most important thing is . . . not to hide behind a professional façade but to come across as a real person, with feelings and thoughts of . . . [one's] own" (Korthagen, 2001, p. 120). Congruence and honesty characterise genuineness, so that to be perceived as genuine, one needs to act in a manner consistent with one's thoughts and feelings (congruence) and be able to express these thoughts and feelings to others (honesty) (ibid, 2001). Allender (2001) highlighted the important difference between the teacher educator enacting the responsibilities of his/her role and the teacher educator making real, connected contact with others.

Revealing myself as a 'real person' created dilemmas for me about what thoughts and feelings I might share with prospective teachers that would be useful for them and, that would support their growth yet not undermine their confidence in me; or in themselves. For example, at times I found it difficult to understand and/or accept ideas expressed by the prospective teachers in the Biology methods class, yet I hesitated to reveal my lack of understanding, to publicly persist in resolving my confusion, or to disagree with a viewpoint. Instead I pretended to understand or agree when I did not; a behavioural response that surfaced under pressure when I felt uncomfortable because I did not want to compromise my supposed expert status as a Biology teacher educator. And, linked to this I also did not want to embarrass individuals nor jeopardise my relationship with them.

Recognising such moments as a teacher educator involves developing one's self-awareness. This includes sensitivity to the interpersonal demands of various situations and a concentration on personal decision-making about that which might be helpful to highlight for students (or not) in a given situation, how to highlight a particular issue/concern/practice/thinking in a given situation, and how students might interpret the teacher educator's responses.

Taking risks, exposing one's vulnerability

Risk taking is essential to genuineness. Stepping outside the boundaries of expected teacher educator behaviours incurs risk yet at the same time offers new opportunities for learning. As indicated in the previous point, my readiness to take risks was moderated by the ways in which I thought prospective Biology teachers might perceive me – although my usual tendency was to plunge 'head first' into a potentially risky situation then realise part way through the risk-potential associated with my actions.

This raises the issue of risk taking as a pedagogically purposeful activity. If prospective teachers are to recognise the meaning and value of their teacher educators taking risks and exposing their vulnerability (so that prospective teachers might consider doing this themselves) then, it is important that they recognise (although not necessarily agree with) the purpose for so doing.

Trusting in oneself and one's students

A person's willingness to take risks in his/her learning grows out of a trusting relationship between members of the learning community. At the same time, there is a need to learn to recognise limits to risk-taking imposed by the context or one's own personality. Loughran (1996) identified from the study of his own teaching practice, the importance of establishing mutual trust between himself and prospective teachers in his classes, in order that they would view his teaching approach as 'purposeful' rather than 'peculiar'. Loughran found that establishing and maintaining trust in his teaching approach was made more difficult when prospective teachers had strongly held beliefs that they could be told how to teach (Loughran, 1996, p. 434).

Trust is linked to acceptance of oneself and others and is demonstrated by paying careful attention to what others have to say, practising 'wait time' and 'withholding judgment'. When I worried about controlling a teaching/learning situation, for instance, in taking over post lesson discussions or steering the learning within a session towards a goal that I thought was worthwhile, rather than waiting and accepting the students' efforts and differing agenda, my capacity for trust was diminished. However, my trust was in evidence when I stood back and supported Lisa's efforts to learn about teaching, herself. Trusting in the capacities of one's students as adult learners who are, ". . . fully capable of making reasonable judgements about . . . [their] own learning and the direction of that learning" (Bullough, 1997, p. 21) allows new ways of seeing to emerge for both prospective teacher and teacher educator alike.

As a teacher educator, I needed to trust that my students viewed my approach to teaching about teaching as one that could provide a stimulus for their exploration of the teaching/learning relationship. Teacher educator trust is important so that prospective teachers can, and will, be receptive to opportunities for learning about teaching.

Compassionate Teaching

These attitudes, when considered together, contribute to what I am calling compassionate teaching. Teaching compassionately means recognizing and letting go of one's own needs and entering into a relationship with others (Jersild, 1955). For me, as a teacher educator, this means learning to suspend my own needs and concerns and entering into the meaning of a situation as another is experiencing it. This is a difficult concept to explain, and an equally difficult concept to 'live' as I grow in my practice. Developing these particular attitudes helps engender a sense of teaching about teaching that goes beyond the simple delivery of ideas, information and theories about teaching and helps to create a bridge into the world of learning through experience; so crucial to helping learners of teaching develop their knowledge of practice in meaningful ways.

Compassion is intimately tied to an understanding and acceptance of oneself and others. Self-study (Hamilton, 1998) is an important means of developing the self-understanding and self-acceptance that I believe is crucial to informing the development of compassionate teaching. Learning about teaching about teaching through self-study has highlighted for me the need to distinguish between aspects of teaching that are problematic in my practice as well as that which is problematic for prospective Biology teachers; an important and defining difference which is regularly played out in the tensions of teaching about teaching. Teaching compassionately therefore requires humility. That is, the ability to wait, to listen, and to withhold judgment about oneself and others (Jersild, 1955).

> The humble person can tolerate himself [sic] not only as one whose knowledge is imperfect but also as one who is imperfect. Here humility interweaves with compassion and provides a person with the beginning of wisdom. It is only when he

can tolerate himself as an imperfect creature, without feeling apologetic about it, that he can have the freedom to listen and learn. (p. 99)

Jersild proposed that humility can be achieved when a person learns not to put unrealistic expectations on him/herself, which in turn: "frees a person to know that so much in life is uncertain, untested, untried and unknown" (ibid, p. 98).

Through this self-study research, I have come to recognise that fears and doubts about my competence as a teacher educator have hampered my capacity to act with compassion and humility. The considerable demands I placed on myself to be able to immediately understand and competently respond to new teaching and learning situations led to feelings of guilt and doubt when I was unable to satisfy my self-imposed expectations. The development of professional self-understanding through analysis of my experience has led to the conceptualisation of knowledge of practice as tensions. Changing the frame of reference, or "reframing" (Schön, 1987) practice in this manner is personally and professionally empowering since it guides new understandings of practice and opens up new possibilities to think and act differently within practice. Self-study of teacher education practices formalises the reframing process (Hamilton, 1998).

EPISTEME, PHRONESIS AND THE 'TENSIONS OF PRACTICE'

Korthagen & Kessels, (1999) proposed that knowledge of practice developed and understood from and through experience "is more perceptual than conceptual" in its nature (p. 7) and encompasses attitudes, feelings, values, thoughts, needs, conceptions, etc. This type of knowledge is called phronesis, or "theory with a small t" (ibid, p. 7). Knowledge as phronesis contrasts with traditional conceptions of knowledge as episteme, ("theory with a big T"), expert knowledge on a particular problem connected to a scientific understanding of that problem. Episteme is propositional (i.e., consists of a set of assertions) that apply generally to many different situations and is frequently formulated in abstract terms. Phronesis, on the other hand, is situation-specific, focuses on strengthening one's awareness of the characteristics of that situation and finding a helpful course of action through it. Hence, conceptualizing the complexities associated with the development of my teacher educator pedagogy through the notion of tensions could be considered an example of phronesis.

Conceptualising tensions and using them as a sign-post for learning to understand and articulate approaches to teaching and learning about teaching helps to highlight the relationship between episteme and phronesis. For example, consider the tension which I experienced in two ways, that of *telling and growth:* between informing and creating opportunities to reflect and self-direct; and, between acknowledging student teachers' needs and concerns and challenging them to grow beyond their immediate preoccupations.

This tension hinges on an acceptance that *telling* is most commonly an attempt to transfer propositional knowledge (that which might apply generally to many different situations, i.e., episteme) from the teacher to the student and, that although such transfer may occur, it does not carry sufficient understanding to the receiver of the information to necessarily be personally meaningful. This tension produced a dilemma for me as teacher educator. It may have been clear to me what prospective Biology teachers needed to know, but this was very different from me knowing how to act in a given situation. Hence I struggled between informing (delivering the propositional knowledge) and creating opportunities for them to reflect and self-direct (making experiences about the issues personally meaningful).

This tension was exacerbated for me by moderating between acknowledging prospective teachers' needs and concerns and challenging them to grow. Interestingly, just as the prospective teachers in my Biology methods class may not have been helped by the delivery of episteme but rather needed to develop their understanding through phronesis, so too I faced the same difficulty in my learning of teaching about teaching – which further highlights the problematic nature of teaching about teaching (i.e., in many situations, developing one's understanding through phronesis rather than through epistemic categorization is necessary). And, as I experienced it, simply knowing about something conceptually, was not the same as doing it, practically. Working from real situations and examining *how* I taught as opposed to *what* I taught, and encouraging prospective Biology teachers to do the same, led to the development of perceptual knowledge for all participants.

Considering this tension in this way highlights an important issue pertaining to experience. The development of knowledge as phronesis is rooted in experience, since it is through systematic reflection on real situations with all their accompanying thoughts, feelings, needs, concepts, etc. that enables an individual to begin to build greater self-awareness and articulate new frames for practice; it was certainly the case for me. For both the prospective teacher and teacher educator alike, this presents a complex and difficult task since inexperience in the role of teacher makes it difficult to know how to act yet it is only through personally involving oneself in teaching/learning situations, that the development of informed action is possible. (This is representative of the Meno Paradox. The classic formulation of the paradox is a dilemma: how can you search for knowledge of something, if you do not know what it is? If you know what it is, then you have already got knowledge of it and cannot search for it. If you do not know what it is, you cannot search for it, because you do not know what you are searching for.) For externally oriented learners, such as myself, (and a number of the students) whose 'default' learning style is to seek guidelines and structure in the form of epistemic knowledge, learning by acting, and developing knowledge of practice as phronesis is potentially confronting and risky, and exacerbates feelings of vulnerability in the learner.

The knowledge developed through this study also shows characteristics of episteme. I now turn to an examination of the relationship between knowledge conceptualised through this self-study as phronesis and, as episteme.

The Phronesis/Episteme Relationship and Knowledge Developed through this Study

In the process of examining what guided my behaviour as a teacher educator and the dilemmas I faced in teaching about teaching, I became more consciously aware of the specific features of particular teaching/learning situations, i.e., I developed my phronesis with respect to these aspects of my teacher education practice. As I articulated and organised these experiences, I named them as 'tensions'. As a consequence of labelling the series of tensions, I was able to explore them in more detail from within my practice and to link them with formal theories (episteme) with which I was already familiar (for example, wait time, or conceptual change). This meant that my knowledge was further developed both perceptually and conceptually as theory and practice informed each other.

My understanding of practice as tensions, initially tied to specific experiences, gradually became detached from those experiences, and abstracted as knowledge in the form of episteme, as I was able to apply it more generally in other (similar) situations. Korthagen (2001) proposes that knowledge at this level is self-evident to the practitioner and can be used in less conscious, more intuitive ways, which means that the individual is freed to concentrate on other things. In this research, conceptualising knowledge as tensions freed me to explore new aspects of the tensions; one outcome of this was my recognition of the role of teacher educator attitudes operating within and across tensions; as described above.

Korthagen (2001) draws on notions from gestalt psychology in order to elaborate a theoretical basis for explaining this process of knowledge development, moving from unconscious behavioural patterns (gestalts), to developing conscious awareness of one's gestalts (developing interiority; level reduction), to knowledge that becomes abstracted from these specific experiences and more available for use in other related situations (schematisation and theory formation), leading to a form of knowledge that requires less conscious attention, leaving the individual able to focus her/his attention on new things, including new aspects of familiar practice (level reduction; new gestalt). Fig. 11.1 (below) summarises the development of learning through experience.

The model shown in Fig. 11.1 suggests that phronesis becomes more elaborated as an individual systematically reflects on the features of a situation. As a consequence, knowledge in the form of phronesis becomes more available both to the practitioner him/herself for further refinement and, when abstracted (as episteme), to others. Korthagen and Kessels (1996) propose that with respect to this knowledge relationship: (i) phronesis is to be considered of higher quality if it is fed by

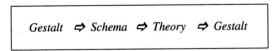

Figure 11.1. Process of Knowledge Development through Experience

episteme; and, (ii) episteme that is not connected to existing phronesis will not effect much change in the practitioner.

It is important to note that conceptualising knowledge as episteme or phronesis is not intended to present a knowledge hierarchy. Episteme is not a higher form of knowledge than phronesis. What is important is how they differ in origin and purpose, and when and how each can be used to better understand the process of knowledge development in teaching/learning situations.

Self Study and Phronesis

This extensive example of self-study research illustrates broadly: (i) the development of personal perceptions while trying to (act to) improve one's own teacher education practices; and, (ii) the results of personal efforts to take research based findings and enact them in personal practice.

The first point relates to the development of phronesis and the second is about episteme informing phronesis (*T. Russell, personal communication, 3/12/2002*). In pursuing (i) and (ii) practice is improved through reframing; that is drawing insights from, reconceptualizing and enacting one's ideas about what it means to be a teacher educator. As a consequence actions and intent become more closely aligned. A further point relates to the communication of this knowledge beyond the individual, that is, (iii) the development of personal perceptions (phronesis) extending and informing research based findings (episteme). This highlights the complementary nature of episteme and phronesis, and the importance of both in knowledge production and dissemination about teacher education practices. In self-study research this point is vital.

The efforts of those engaged in self-study research must inform both the individual and the community of teacher educators. That is, "the commitment to provide insights for others of how the understandings of the authors become part of their actual day-to-day practice" (Hamilton & Pinnegar, 1998b, p. 242). Making knowledge available to others as phronesis is an important task in establishing and building a professional knowledge base of a pedagogy of teacher education. Both episteme and phronesis are forms of knowledge that need to be developed and made available both within the practice of individual teacher educators and their students and across the community of teacher educators.

Traditional approaches to knowledge production mean that we (i.e., those working within the academy) already know how to capture and share episteme. The challenge that confronts teacher educators is in finding ways of capturing, portraying and sharing phronesis. A significant challenge for teacher educators engaged in self-study research lies in finding ways of communicating these insights and understanding in ways that are meaningful and useful to other teacher educators.

Knowledge of practice as phronesis offers one way of communicating the holistic nature of experience. This self-study is an attempt to address this challenge through representing knowledge of practice in the form of phronesis through the vignettes of practice portrayed in this section, and that might be accessible for use and exchange within the community of teacher educators.

SUMMARY

To summarise, this chapter has reviewed and made explicit interconnections between the set of tensions of teaching about teaching that I identified from within my teacher education practices. The chapter also illustrates the holistic nature of the tensions and the ways in which knowledge of practice as tensions, emerging from my analysis of practice, illustrates knowledge both in the form of episteme and phronesis.

In self-study, it is important to articulate the knowledge developed through studying practice and, how knowledge developed through researching practice impacts practice. In the next section, I describe the ways in which this self-study has impacted my everyday teacher education practices and how my work as a teacher educator continues to be shaped by my self-study research.

PART THREE

LEARNING FROM TEACHING ABOUT TEACHING

Chapter Twelve

BECOMING A TEACHER EDUCATOR

[T]eaching is a life-long learning process . . . one doesn't eventually become a teacher [educator]; but instead moves in understanding teaching/learning through active involvement in the process. (Guilfoyle, 1995, p. 18)

INTRODUCTION

In self-study, it is important that the knowledge developed through the study of one's practice, informs practice. In so doing, the learning from self-study impacts at a personal level, as a starting point for influencing one's own practice and the practice of others (Loughran, 2004). Self-study can therefore be conceptualised as an ongoing spiral of research informing practice, informing research. In this final section of the book, I outline the ways in which my work as a teacher educator has been, and continues to be, shaped by the knowledge developed (and continuing to develop) through my ongoing self-study. I illustrate the nature of my learning about practice and how this has impacted my understanding of the complexities associated with becoming a teacher educator.

LEARNING FROM TEACHING ABOUT TEACHING

When I began this research project I anticipated that the process of studying my practice as I engaged in it would position me well to generate, and then implement, new insights and understandings of teaching and learning about teaching with prospective teachers in Biology methods classes. My immediate needs were concerned with learning how I could better teach about teaching Biology so that the prospective teachers in my classes could be better teachers of Biology. However, a continuing difficulty that I faced as I began to investigate my practice was that the kinds of insights and understandings that I sought did not become immediately apparent in the manner I had expected. In fact, the deeper I began to probe into the teaching/learning relationship, the more complex and problematic were the issues that I encountered. As a practitioner/researcher I regularly faced situations in my daily practice that I could not understand or solve at the time and that, as a consequence, led me to feel frustrated and 'stuck'.

Other educators researching their practice have felt this, too. For example, Northfield (Loughran & Northfield, 1996) described the frustration associated with his efforts to better understand and improve his classroom teaching, as he practiced it: " . . . my frustration during the year was in trying to analyse the day to day teaching experiences in a way that might lead to consistent improvement in classroom interactions . . . feeling like I should have been able to better understand and use my experience" (p. 135). As a consequence of his study, Northfield came to recognize that the way in which his knowledge was developed and drawn upon as a teacher, was different from other forms of knowledge about teaching (for example, 'T'heories about teaching). So too, in this self-study research, I have come to see that the knowledge of teaching about teaching that I sought, and the nature of the knowledge that I have developed, including the ways in which this knowledge has emerged, has grown and changed throughout the research process.

The manner in which I conceive of my learning over the time of this self-study can be organized according to four areas:

1. Reflecting on the process of engaging in this research
2. Understanding the nature of self-study
3. Knowledge of practice developed through this research
4. 'Tensions' as a conceptual frame for doing and understanding research.

Reflecting on the Process of Engaging in this Research

One of my initial goals in framing this study was to consider the alignment between my actions and intentions and to uncover ways in which various types of assumptions I held, played out in my pedagogy. As a beginning point for the process of "assumption hunting" (Brookfield, 1995) I posed a series of questions about my pedagogy that I wished to address through this self-study. These questions emerged through my analysis of experience as a learner and as a teacher. At the time of writing them, I conceptualised the questions around the broad frame offered by self-study: "How do I improve my practice? How do I live my values more fully in my practice? How do I help my pupils to improve the quality of their learning?" (Whitehead, 1993, p. 113).

The act of writing about and researching my practice using this broad frame began to stimulate a process of change in my understanding of practice and consequently, my approach to research. For instance, noticing aspects of my practice as problematic, rather than 'simply' as a series of problems meant that I was then able to probe those problematic situations more fully in order to learn about how they impacted the teaching/learning relationship. However, it was not until I was well into the formal research process that I began to conceptualise practice as managing a set of competing tensions.

As I engaged in the process of analyzing my practice as tensions, I came to better understand my knowledge of each tension and therefore better able to articulate it. Some tensions are less well elaborated in their representations of the knowledge of practice that I developed compared with others, which also represents how I became more confident in my understanding of the process of analysis and more familiar with themes and patterns that were emerging through this newly conceptualized frame. Hence, this self-study is itself evidence of how I took myself to new levels of understanding about, and articulation of, my practice.

In order to illustrate for the reader part of this process of my knowledge development, I return to the questions I posed at the outset of this study (see Chapter 1) to consider how each has been addressed and reconceptualised through this self-study, 'reading' them through the newly developed frame of 'tensions'. For example, one of the questions I posed initially, was: *How can I create a methods course that acknowledges prospective teachers' histories as learners and is responsive to their needs, yet at the same time challenges their views and gives them the confidence and reason to try alternative approaches to teaching senior Biology (particularly when I have never experienced such a methods course myself)?* This question can now be understood as exploring issues of acknowledging prospective teachers' needs and concerns and challenging them to grow; that is framed in the tension of 'telling and growth'. It may also be read through the tension of 'safety and challenge', between a constructive learning experience and an uncomfortable learning experience. Of course, when I first articulated these questions about my practice, I was unaware of the tensions embedded within them. They were 'simply' aspects of my pedagogy that I wanted to learn more about. However, in revisiting these questions now, I can identify the various tensions encapsulated within them, and in addressing my questions about my practice I recognize that they are a part of larger concerns that I face in teaching about teaching.

Using the tensions as a frame for understanding my questions helps me to be more purposeful and focused in the way I address them. For instance, in answering the question about how I might create a methods course that both acknowledges and challenges the experiences of prospective Biology teachers, I can now say that the teacher educator's attitude matters (working compassionately with students as individuals to support and encourage them – and recognizing this as practice not just rhetoric), so too does selecting experiences for prospective teachers that challenge their thinking and offers them opportunities to try out new practices, as well as monitoring the effects of the learning by talking about these experiences (over time) in focused and purposeful ways, including difficulties faced by both the teacher educator and the prospective teachers.

For the teacher educator these aspects come together as modeling a 'way of being' as a teacher. Further to this, what I previously conceived of as a need to give prospective teachers, that is, the 'tips and tricks' of Biology teaching, I now see as a point of growth to viewing the teacher educator role as one of creating opportunities for prospective teachers to direct their own learning and development so that it is personally meaningful and effective in facilitating their students' Biology learning. This relates to the tension of 'telling and growth'.

Table 12.1 has been constructed to offer an over view of the questions that I posed about my practice, together with the tensions I can now 'read' within them. It also includes a brief explanation of how I have addressed each question within my practice in order to offer a brief glimpse of 'what was' at the outset of this study as well as 'what is' at its completion.

Table 12.1. Reframing Questions about Practice through Tensions

My Original Question about Practice	Tension that 'Recognises' this Question and Brief Explanation of the Tension	Implications for Practice
Qu: What explicit and implicit messages about learners and learning do I convey to others through the manner in which I conduct Biology methods classes? *Are these messages consistent with those that I wish to develop in my students?*	**Action & intent:** Exploring discrepancies between goals that teacher educators set out to achieve through their work and ways in which these goals can be inadvertently undermined by the way they work.	Being aware of this tension helps me to operate more mindfully within it. I have learnt that not all students interpret my actions in the same way. I have learnt to be less defensive and accept that not all students will accept or find helpful my approach. I now regularly gather feedback from students about their experiences of learning in my classes.
Qu: Which groups of students do I interact with most often or most successfully? *Which groups of students do I marginalise because I do not understand where they are coming from?*	**Safety & challenge:** Engaging students in forms of pedagogy that challenge and confront. **Telling & growth:** Between acknowledging student teachers' needs and concerns and challenging them to grow beyond these immediate concerns	I have identified that those students most similar to me are those whom I tend to interact with more successfully. I recognise that previously I have tried to impose my views on students rather than genuinely seeking their perspectives. By increasing wait time, delaying judgment, and listening to my students I can support and challenge more effectively.
Qu: How much does it matter to me that students like me? *What assumptions do I now make about how prospective teachers approach their learning in my classes?*	**Safety & challenge:** Dealing with encounters that may challenge our relationships with others. **Planning & being responsive:** Having an agenda based on assumptions about learning and being responsive to needs of individuals and particular situations as they occur.	My increased self-awareness means that I do not deny these feelings but become aware of how they impact my interaction with students and consciously learn to shift focus from me to the students.

Qu: How can I create a methods course that acknowledges prospective teachers' histories as learners and is responsive to their needs, yet at the same time challenges their views and gives them the confidence and reason to try alternative approaches to teaching senior Biology (particularly when I have never experienced such a methods course myself)?	**Telling & growth:** Balancing learners' needs and concerns and challenging them to grow beyond their immediate preoccupations.	As part of their work students examine and re examine their learning past. As part of this process I support them as they learn to articulate their needs, and offer supported experiences that still take them out of comfort zone.
Qu: How can I support and encourage students to risk using approaches that would not be considered the 'traditional science' that they experienced and enjoyed? *Why would they want to?*	**Safety & challenge:** Recognising differences between a constructive learning experience and an experience that causes students to disengage due to personal discomfort. **Confidence & uncertainty:** Making explicit my uncertainties in ways that help student teachers see into my pedagogy, yet maintains their confidence in me and their ability to make progress.	Being open about taking risks in my own practice and making learning the focus not my feelings of vulnerability. This means recognising then 'letting go' of the emotions associated with the experience in order to help others' learning.
Qu: How can I sensitively monitor the feelings of the prospective teachers that I work with so that they might be encouraged to participate in experiences that may seem uncomfortable? How can I help prospective teachers recognise that the feelings of the learners that they work with in Biology classrooms will	**Safety & challenge:** there is a difference between experiences that may be 'safe' and acceptable to me and feelings of safety acceptable to student teachers; this is different for different individuals. **Confidence & uncertainty:** by making explicit the complexities and messiness of teaching and at the same time helping	Importance of cultivating attitudes that accept that risk taking is an important and worthwhile aspect of learning about teaching. By encouraging prospective teachers to put themselves 'in the shoes of others' and imagine their responses to different

(Continued)

Table 12.1 (Continued)

My Original Question about Practice	Tension that 'Recognises' this Question and Brief Explanation of the Tension	Implications for Practice
influence their participation and what can be learnt?	student teachers feel confident to recognize that these complexities are an important and manageable aspect of teaching.	situations in order to create new understandings of practice.
Qu: How can I help prospective teachers learn to trust that they can learn from each other and that my knowledge and experience may not work for them in their practice? *And, how can I continue to see my practice through others' eyes so that I might remain sensitive to others' needs and concerns?*	**Confidence & uncertainty:** Balancing student teacher needs to see me as a leader and helping them develop confidence in themselves and each other.	Through giving prospective teachers a chance to try new approaches and ideas, and helping them to see why and how a particular approach might be used, within a supportive learning environment. Continue researching my practice, seek to find ways to more deeply explore the others' experiences of my teaching.
Qu: If I have rejected an information transfer model of learning as unrepresentative of reality, how do I deal with prospective teachers who think about learning in the way that I did? How do I know what models of learning the prospective teachers in my class hold, and what are the effects of their views of learning on the way they interpret my teaching?	**Telling & growth:** between acknowledging that prospective teachers bring particular views about teaching and learning and creating opportunities to reframe these views in the light of ongoing experiences. **Acknowledging & building on experience:** between being mindful of prospective teachers' prior experiences and ideas, but not reinforcing these by the approach taken.	Work actively towards developing a stronger sense of empathy with needs of others. Develop new ways, further refine existing of learning more about those I work with, and their motivations. More strongly developed sense of the meaning of 'wait time'. Trusting in the value of the approach that I take and as part of this, remembering that learning happens over time, so also trusting that this process will occur.

Question	Tension	Response
Qu: How do I account for those prospective teachers who find my approach to learning personally threatening, since it involves exposing and challenging their "untested thinking" in a public forum?	**Safety & challenge:** between creating opportunities for learning that engage and invite new approaches to practice and preventing student teachers from learning because the challenge is too great. **Confidence & uncertainty:** between helping prosepctive teachers to try out new ideas and learning to critique these ideas without feeling defensive.	Greater modeling in my practice of my thinking about practice and encouraging prospective teachers to begin to do this for themselves. Developing greater sensitivity to the particular needs of individuals in my classes through genuinely 'listening with' them.
Qu: Although I am philosophically aligned to a particular way of thinking about learning, are my teaching actions consistent with this?	**Action & intent:** between working towards an 'ideal' and being mindful of the approach taken to achieve this 'ideal'	Continue to investigate my practice. There will be no certain answer to this but instead, it will always be a 'moving towards' in the process of becoming a teacher educator.

Understanding the Nature of Self-Study

Self-study is a form of practitioner research whereby the context for the research is practice itself. It is also a powerful tool for uncovering important facets of the knowledge of practice (Loughran, 2005). The experience of conducting this research as a longitudinal self-study has further enhanced my understanding of the features of self-study as an approach to researching practice, and has led me to new insights about the nature of self-study as a way of understanding the work of the teacher educator through bringing together teaching about teaching *as practice* and teaching about teaching *as research*. A discussion of each of these aspects now follows.

Better understanding the features of self-study

- *Self-study* as *'responsive' research:* Self-study is a form of research that is responsive to the demands of the practice context. This means that teaching and research become intertwined as new experiences and insights derived from the teaching context lead to shifts in the research focus and in turn, give rise to other insights, possibilities and actions so that over time practice and research focus are constantly changing in response to the changing context (Loughran, 1999). In this research, the questions that initially framed my self-study led me to recognize new aspects of practice. I would assert that this is where a self-study methodology is different to more traditional research approaches and where doing self-study is important to understanding the genuine power of self-study. The insights available from the research process can impact learning about teaching and research rather than simply producing knowledge outcomes of an epistemic nature. This links to the next point as the shifting focus of the research lens means that:
- *Learning can be slow to emerge through self-study:* The construction of knowledge about practice through self-study is often experienced as a continuous and evolutionary process (Kremer-Hayen & Zuzovsky, 1995). Throughout the process of acquiring and examining experiences of teaching about teaching, I was developing new levels of understanding about my practice and learning to articulate my knowledge in new ways. As part of this process I needed time to understand my practice, to recognize the need for change, and then choose (or not) to alter my behavior. Uncovering differences between my actions in practice and my intentions for practice has been a slow and complex process because my habits of practice are so deeply ingrained.

  Self-study researchers recognize that deeply held (paradigmatic) assumptions are hard to uncover, resistant to examination and require considerable (disconfirming) evidence to overturn (Brookfield, 1995). My experience of change as "evolutionary rather than revolutionary" (Fullan, 1991) reminds me that prospective teachers also need time and suitable opportunities to begin to be aware of, and act upon, their own needs for change.

Numerous small shifts took place (and continue to take place) within my practice because the teacher (educator) has a responsive role. Many of these changes are already reported in the previous section of this book in the ways in which I thought about and approached the teaching and learning about teaching in Biology methods classes. Other changes either went unnoticed by me (because of the tacit nature of some aspects of practice), or were unable to be articulated, because although I was aware of a change in my practice, at that time I was not yet able to name it. The particular contextual demands of individual Biology methods classes influenced the extent and type of change that took place in my practice. The cumulative effect of these various changes led to my conceptualisation of practice as managing a series of tensions.

- *Self-study is self-focused but not self-centered:* As self-study researchers acknowledge, there is an inherent ambiguity in the label self-study. One's self is (at least initially) the focus of the study, but while self is the focus, the focus on self can also be limiting if one cannot see that the reason for studying self is to develop one's perceptions of how self is experienced by others (Bullough & Pinnegar, 2004). Because the researcher is located at the centre, albeit with a purpose of modeling practice in such a way that beliefs and actions are made more congruent, then a danger emerges that the researcher becomes the focus of the research as new understandings of self are revealed; understandings that can be confronting, surprising and induce vulnerability.

  As I engaged in the research process, I found it seductive to become caught up in my self as the purpose and focus. The longitudinal and personal nature of this study meant that at times, the focus on me, and my feelings, compared with those of the Biology method students, was inevitable. Sometimes however, I became trapped in the feelings of vulnerability and guilt that a sustained examination of one's teaching practice can quickly induce. As noted by Bullough and Pinnegar (2004) this can raise problems for the teacher educator as a person (I needed self-affirmation to continue) and as a self-study researcher (for me, in deciding what to make public from this research). As a consequence of dealing with these problems, I became more aware of the importance of knowing about the ways in which self-study influences one's self, for instance by monitoring where the focus of the study lay at any particular time (i.e., whose feelings are at 'centre-stage' and why?). This helped me to maintain a sense of purpose and, when necessary, to refocus my efforts in order to enhance my teacher educator practice so that there might be similar outcomes in relation to my students' learning about teaching.

- *Self-study is demanding work:* Maintaining a sustained focus on one's work through the ongoing analysis of practice is a difficult task. At times, in conducting this research, I wanted to escape from the intensity of the process to give myself a rest. However, I learnt that taking 'time out' does not mean periodically abandoning self-study, (as one might decide to stop and do some other form of research for a while). Instead, it means being kind to oneself through the process of researching practice – for example, using approaches to teaching that one enjoys, knows are usually fruitful and that students enjoy, and not subjecting every movement and thought to detailed critique.

While my understanding of the features of self-study became more elaborated throughout the period of this research, an insight into self-study that emerged through the research and that is new for me, relates to the way in which the work of teacher educators can be reconceptualised through self-study.

Self-study as a means of better understanding the work of the teacher educator

Traditional notions of teacher educators' work separate the worlds of teaching and of research. Similarly, traditional notions of research separate the researcher from the researched. Depending on one's perspective, teaching is what teacher educators do when they are not doing research (or some other Faculty related activity), or what teacher educators should focus on and do 'best' because of their important role in facilitating the learning of others. Self-study brings together the acts of teaching about teaching, and research about teaching, within the particular context of the individual teacher educator via activity that directs the development of both practice and research. Better understanding the knowledge of practice is integral to the development of the work of each individual teacher educator, and for teacher educators as a collective. The development of this research project has led to my understanding of self-study as uniquely positioned to support such a process of personal and public knowledge development.

Knowledge of Practice Developed through this Research

The following assertions encapsulate the knowledge of practice that I have developed through this self-study of my practice. They are:

• *Framing and reframing is now central to my decision-making* as a teacher educator. Taken-for-granted views of teaching and learning act as frames for interpreting practice (Schön, 1983). "Reframing" (Schön, 1983) involves recognizing alternative perspectives and approaches in learning situations. The experience of developing my understanding of practice as tensions was a very useful experience of reframing, helping me make sense of my practice in new ways and, in the process, leading me to a deeper understanding of my teaching and learning about teaching. Reframing is important to the process of self-study as reframing acts as a mediating factor in decision-making, influencing responses and actions. Viewing practice now through the new frame of tensions enables me to maintain a sense of purpose while teaching and, at the same time, offers an advance organizer for thinking about, and communicating, practice to a wider audience (Mitchell, 1999). Reframing also led me to new understandings of what it means to be a teacher educator, for example, in coming to recognize the shift from

seeing my role as offering a supply of good ideas about Biology teaching to seeing learning about teaching as being problematic. Being able to see how these
different aspects impact on learning about teaching is linked to the next point, as:

- *Making a distinction between technical aspects of practice and understanding of practice* is now helpful to the integrated development of my learning about teaching, as opposed to maintaining a view of these aspects as dichotomies. I now see
that technical aspects of teaching, such as different teaching approaches and activities, function as tools for supporting the development of prospective teachers'
understandings about teaching and their confidence in feeling better prepared as
teachers, rather than viewing them as ends in themselves. The teacher educator's
role lies in constructing experiences that lead to reflection on the use of these tools
and the contexts in which they are, and might be, used so that prospective teachers
may be better able to do this for themselves in their future teaching.

- *Responding to and balancing prospective teachers' needs and concerns* is important in informing my knowledge of what will help them progress as learners of
teaching. This is closely linked to the previous point in so far as the teacher educator adopts a responsive role, recognizes differences between short term and
long term needs of prospective teachers and identifies these needs, not as a
dichotomy, (either I address short term, or long term) but instead, embracing both
simultaneously. One of the important aspects of my learning from this self-study
is recognizing that binaries can co-exist; the challenge for me lies in learning to
work towards balancing the elements associated with each.

- *Trusting prospective teachers to learn about teaching for themselves is essential.*
This means believing in the capacity and motivation of students to take responsibility for making their own meaning and progress in learning about teaching,
acknowledging that this occurs for individuals in different ways and, that it is a
process facilitated by teacher educator encouragement and support. I learnt that
the more I acted in a supporting role (compared with a controlling and/or critical
role) the more active students became in their own learning. This is not to imply
that the teacher educator does nothing other than provide friendly encouragement, rather it means guiding students' learning while at the same time being
respectful of each learner's right to direct his or her own self-development.

- *Learning about my 'self' through developing my self-understanding and self-
awareness* is prerequisite to helping others see themselves in ways that enable
them to help themselves (just as I have learnt to do). Exploring my biography as
a teacher and learner as a prelude to this study enabled me to identify myself as a
"received knower". As a consequence, I developed more clarity about my difficulties in with dealing with the uncertainties of dilemmas and began to listen to,
and trust, my own voice. Identifying the tensions helped me understand myself as
a teacher educator struggling with dilemmas. Hence, reframing my pedagogy as
a series of tensions to be managed is evidence of having moved beyond a
"received knower" approach and towards deeper levels of self-understanding.
Through this self-study, I have also come to recognize some of the ways in which
my self-growth has been stifled by previous concrete and dualistic ways of

thinking and acting; that does not mean the situation is changed forever, but more so, that I more readily recognize how I am functioning within a given situation and have more control and confidence in how I respond.

Feelings of anxiety, lack of confidence and perceiving situations as "either–or" binaries are also experienced by prospective teachers as they prepare to face new and challenging situations. Through learning about these feelings for myself I have become more sensitive to the ways in which such feelings manifest and stifle growth in prospective teachers. Although I strived to do so previously, I have learnt that there is an important difference between being (and appearing) vulnerable to one's own emotions and finding ways of disclosing one's emotional state so as to facilitate learning for one's self and others. Growth in self-understanding has led to growth in my confidence as a teacher educator and in my relationships with my prospective teachers.

- *There are recognisable shifts in "becoming a teacher educator".* Accepting the title of teacher educator does not bring with it knowledge of how to act in the role. A transition in one's understanding occurs when the teacher educator is prepared to view practice as problematic and to look beyond the transmission of 'good ideas about teaching'. Recognising the inadequacy of the "telling, showing, and guided practice approach" (Myers, 2002) leads to a changed view of learning about teaching about teaching which then impacts practice and begins the process of the development of new knowledge both for oneself and one's students. For me, evidence that this shift was occurring was recognizable in aspects such as knowing more about prospective Biology teachers' needs and concerns and the ways in which I responded to them, becoming more sensitive to how and when to challenge these new teachers' beliefs about teaching and learning and, how I could help individual students grow in ways that might be self-actualising for themselves and their students. Taking changed thinking into practice and drawing on new sensitivities about teaching and learning about teaching marks a further shift in teacher educator development. Viewed in this way the notion of 'becoming' a teacher educator is experienced as a continuing process. The work of the teacher educator is in the development and refinement of his/her phronesis and in making the knowledge developed as phronesis explicit and available to others (i.e., self, students, colleagues, research community) in ways that are useful and useable for that group.
- *Collaboration* (in lots of ways) leads to being challenged about taken-for-granted assumptions and helps build knowledge of practice. Through the shared experiences of Biology methods classes (with prospective teachers and a colleague) I was able to gain a range of perspectives on our experiences and develop greater meaning from these experiences because they were shared. Acknowledging the alternative perceptions of different participants then forced me to confront some of my assumptions about practice, leading me to reframe my thinking about my practice. Collaboration also supports the personal demands associated with self-study research since a shared experience can be made much more personally manageable than when that experience is borne alone.

In describing these changes to my practice I can see that the new understandings of my pedagogy that now influence my current practice can be classified as attitudinal changes and structural changes. Attitudinal changes include the ways that I think and feel about my practice and structural changes refer to the ways in which I reorganised the Biology methods program so as to better accommodate my transformed ideas. These responses incorporate concrete shifts in practice as well as conceptual shifts in understanding and hence illustrate the development of my knowledge as phronesis (through perceptual knowledge such as attitudinal change), and as episteme (as I articulated new knowledge of practice and applied it to the reorganization of the curriculum).

'Tensions' as a Conceptual Frame for Doing and Understanding Research

For me, the notion of 'tensions' serves as a frame for conceptualizing the problematic nature of practice, which then helps to formalise the experience of being a teacher educator, and in the process provides a language for articulating knowledge of teaching about teaching.

Development of tensions through the research

One of my initial goals in framing this study was to consider the alignment between my actions and my intentions and to uncover the ways in which various types of assumption (e.g., deeply held, surface) played out in my pedagogy. Consequently, in the data collection phase I was 'assumption hunting'. The notion of tensions grew out of my working through the literature, thinking about the struggles of other teacher educators in their work, concurrently with my own data collection, and thinking about the struggles associated with my teaching about teaching. As this self-study progressed, 'assumption hunting' became subsumed into one of tensions – that of action and intent. This meant that, at least initially, I was most aware of the tension of action and intent within my practice because it was directly connected to the impetus for the study.

Although at the time I did not conceptualise these ideas as tensions, I was preoccupied with the notions of safety and challenge, and of confidence and uncertainty due to such factors as feeling sensitive about conducting a study of my practice (the sense of personal vulnerability is high), the way in which the Biology methods students were interpreting and responding to my chosen pedagogical approach (it was challenging for many of them which caused them to feel threatened and insecure), and my developing ideas about how I 'should' teach in order to support students' meaningful learning about Biology teaching.

As the study progressed these ideas began to take shape as tensions. Reflecting on the research process now, I can see the ways in which each of the tensions came into focus at different times and for different reasons. The extent to which any particular tension was evident was dependent on the mix of contextual factors (my feelings, feelings of different students, time of the year, etc.) operating at any given time.

Also, because there are different 'levels' of complexity within each tension, various aspects of different tensions became more apparent to me at some times compared with others. For instance, in some situations my preoccupation with 'surviving' meant that for some time, my understanding of the tension between *safety and challenge* only really existed at a superficial level. It was not until later, after careful analysis of the different sources of data from my practice about this tension (e.g., finding differences between my conceptions of safety and what my students considered as 'safe'), that the deeper, more subtle dimensions of this tension became evident.

As this research progressed, and the tensions emerged as a reasonable frame for understanding pedagogy, I became more aware of instances of a particular tension in my teaching. I began to recognise and name them in my practice. Even so, I recognise that it is not possible to be aware of all of the tensions operating at any one time. In my current practice I find that the six tensions serve as a means of understanding practice, retrospectively. I am not yet able to conceptualise practice as tensions, as I teach. This means that currently, the tensions serve as a way of thinking about practice, afterwards. However, I am moving towards using what I know about the tensions as an 'advance organiser' for practice as I anticipate situations based on my previous experiences.

Through the process of reframing my work as a teacher educator managing tensions, I am also better able to recognise some limitations to the ways in which I previously conceptualised my pedagogy (as personal deficiencies, as romantic professional ideals) and the possibilities for new learning and growth inherent in conceiving practice as managing tensions. The effect of this is that my inclination to reconsider my practice is not only renewed, but given a richer lens through which to look, with new opportunities and inevitably, new difficulties. I experience a sense of personal strength in having a frame for understanding and researching practice that is more robust, more organised and more purposefully directed at teaching and learning about teaching.

The notion of 'tensions' and the set of tensions I have identified offers insight into the work of teacher educators as they seek to facilitate the development of prospective teachers. The tensions then serve both as a frame for studying practice and as a language for describing practice and, in so doing, may be considered a way forward in developing a pedagogy of teacher education that can be shared amongst the community of teacher educators.

Conceptualising tensions and using them as a sign-post for learning to understand and articulate approaches to teaching and learning about teaching might help others see how tensions impact practice and bring into focus approaches to teaching about teaching that might make more tangible the relationship between episteme and phronesis.

The development of knowledge of practice is important for teacher educators in order to improve the quality of teaching about teaching (and their students' teaching). The tensions themselves are not necessary *per se*, but offer a useful way of conceptualising and communicating practice. A series of tensions is not rich or comprehensive enough to provide guidance in all situations – nor would it be expected to. In this case, tensions offer a way of 'reframing' traditional notions of knowledge development in teacher education and so create new and different possibilities for understanding and improving practice. What is helpful (necessary) about the set of tensions is that it captures and holds onto, "ambivalence and contradiction, rather than reducing or resolving it" (Stronach et al., 2002, p. 121).

The research described in this book, as a comprehensive example of self-study, offers a window through which to understand the process of becoming as a teacher educator. This concluding section has attempted to situate that process in the real world of my teaching about teaching and prospective Biology teachers' learning about teaching. In so doing, I hope that my learning through this self-study has been made clear in such a way as that it might also be meaningful and applicable in the work of others; and that is crucial to enhancing the work of teaching and learning about teaching.

REFERENCES

Adler, S. A. (1991) Forming a critical pedagogy in the social studies method class: The use of imaginative literature. In B. R. Tabachnick, & K. M. Zeichner (Eds.), *Issues and practices in inquiry-oriented teacher education* (pp. 77–90). London: Falmer Press.

Adler, S. A. (1993). Teacher Education: Research as reflective practice. *Teaching and Teacher Education*, 9 (2), 159–167.

Allender, J. S. (2001). *Teacher Self: The Practice of Humanistic* Education. Lanham, Md.; Oxford: Rowman & Littlefield.

Aubusson, P. (2006). Columbus and Crew: Making Analogical Reflection Public In P. Aubusson and S. Shuck (Eds.), *Teacher Learning and Development: The Mirror Maze*. Springer.

Ausubel, D. P. (1960). The use of advance organizers in the learning and retention of meaningful verbal material. *Journal of Educational Psychology*, 51, 267–272.

Baird, J. R., & Northfield, J. R. (Eds.), (1992). *Learning from the PEEL Experience*. PEEL Publications, Melbourne: Monash University Printery Services.

Ball, S. J., & Goodson, I. F. (Eds.), (1985). *Teachersí Lives and Careers*, London: Falmer Press.

Bass, L., Anderson-Patton, V., & Allender, J. (2002). Self-Study as a way of teaching and learning: a research collaborative re-analysis of self-study teaching portfolios. In J. Loughran & T. Russell (Eds.), *Improving teacher education practices through self-study* (pp. 56-70). London: Falmer Press.

Barnes, D. (1998). Foreword: Looking Forward: The Concluding Remarks at the Castle Conference. In M. L. Hamilton (Ed.), *Reconceptualizing teaching practice: Self-study in teacher education* (pp. ix–xiv). London: Falmer Press.

Belenky, M. F., Clinchy, B. M., Goldberger, N. R., & Tarule, J. M. (1986). *Women's ways of knowing: The development of self, voice and mind*. New York: Basic Books.

Berry, A., & Loughran, J. J. (2002). Developing an understanding of learning to teach in teacher education. In J. Loughran & T. Russell (Eds.), *Improving teacher education practices through self-study* (pp. 13–29). London: Falmer Press.

Berry, A., & Loughran, J. (2005). Teaching about teaching: the role of self study. In Mitchell, C., Weber, S. & O'Reilly-Scanlon, K. (Eds.), *Just Who Do We Think We Are? Methodologies for autobiography and self-study in teaching.* (pp. 168–180). Routledge Falmer. London and New York.

Berry, A., & Scheele, S. (2007). Professional Learning Together: Building teacher educator knowledge through collaborative research. In A. Berry, A. Clemans & A. Kostogriz (Eds.), *Dimensions of Professional Learning: Professionalism, practice and identity*. Sense Publishers: Rotterdam. The Netherlands.

Brandenburg, R. (2004). Roundtable reflections: (Re) defining the role of the teacher educator and the preservice teacher as 'co-learners'. *Australian Journal of Education*, 48(2), 166–182.

Brandenburg, R. (2007). *Learning and Teaching in Teacher Education: A self-study*. Unpublished doctoral thesis. Melbourne: Monash University.

Britzman, D. P. (1986). Cultural myths in the making of a teacher: Biography and social structure in teacher education. *Harvard Educational Review*, 56, 442–456.

Britzman, D. P. (1991). *Practice makes practice: A critical study of learning to teach*. Albany, NY: State University of New York Press.

Brookfield, S. D. (1995). *Becoming a critically reflective teacher*. San Francisco: Jossey-Bass.

Bullough, R.V. (1996). Practicing theory and theorizing practice in teacher education. *Teacher Education Quarterly*, 23(3), 13–31.

170 REFERENCES

Bullough, R. V. (1997). Practicing Theory and Theorising Practice in Teacher Education. In J. Loughran & T. Russell (Eds.), *Teaching about teaching: Purpose, passion and pedagogy in teacher education* (pp. 13–31). London: Falmer Press.

Bullough, R. V., & Gitlin, A. D. (2001). *Becoming a Student of Teaching. Linking Knowledge Production and Practice*. New York: Routledge.

Bullough, R. V. Jr., & Pinnegar, S, E. (2004). Thinking about the thinking about self-study: An analysis of eight chapters. In J. J. Loughran, M. L. Hamilton, V. K. LaBoskey, & T. Russell (Eds.), *International Handbook of Self-study of Teaching and Teacher Education Practices* (Vol. 1, pp. 517–574). Dordrecht: Kluwer publishing.

Carson, T. R. (1997). Reflection and its resistances. In T. R. Carson & D. Sumara (Eds.), *Action research as a living practice* (pp. 77–91). New York: Peter Lang.

Cochran-Smith, M. (2004). Walking the Road: Race, Diversity, and Social Justice in Teacher Education. New York: Teachers College Press.

Cochran-Smith, M., & Lytle, S. L. (2004). Practitioner Inquiry, Knowledge and University Culture. In J. J. Loughran, M. L. Hamilton, V. K. LaBoskey, & T. Russell (Eds.), *International Handbook of Self-study of Teaching and Teacher Education Practices* (Vol. 1, pp. 602–649). Dordrecht: Kluwer publishing.

Clandinin, D. J. (1995). Still learning to teach. In T. Russell & F. Korthagen (Eds.), *Teachers who teach Teachers* (pp. 25-31). London: Falmer Press.

Cole, A. L., & Knowles, G. (1995). Methods and issues in a life history approach to self-study. In T. Russell & F. Korthagen (Eds.), *Teachers who teach Teachers* (pp. 130–151). London: Falmer Press.

Connelly, F. M., & Clandinin, D. J. (1985). Personal practical knowledge and the modes of knowing: Relevance for teaching and learning. In E. Eisner (Ed.), *Learning and teaching the ways of knowing* (84th yearbook of the National Society for the Study of Education, Part II, pp. 174–198). Chicago: University of Chicago Press.

Conle, C. (1999). Moments of interpretation in the perception and evaluation of teaching. *Teaching and Teacher Education, 15*, 801–814.

Crowe, A & Berry, A. (2007). Teaching Prospective Teachers about Learning to Think like a Teacher: Articulating *our* Principles of Practice. In Tom Russell and John Loughran (Eds.), *Enacting a pedagogy of teacher education*. London: Routledge.

Crowe, A., & Whitlock, T. (1999, April). *The education of teacher educators: A self-study of the professional development of two doctoral students in teacher education*. Paper presented at the meeting of the American Educational Research Association, Montréal, Canada.

Dewey, J. (1933). *How we think: A restatement of the relation of reflective thinking to the educative process*. Boston: D.C. Heath and Company.

Dinkelman, T. (1999, April). *Self-study in teacher education: A means and ends tool for promoting reflective teaching.* Paper presented at the annual meeting of the American Educational Research Association, Montréal, Canada.

Dinkelman, T. (2003). Self-study in teacher education: A means and ends tool for promoting reflective teaching. *Journal of Teacher Education, 54*(1), 6–18.

Dinkelman, T., Margolis, J., & Sikkenga, K. (2001, April). *From teacher to teacher educator: Experiences, expectations, expatriation.* Paper presented at the annual meeting of the American Educational Research Association, Seattle.

Dinkelman, T., Margolis, J., & Sikkenga, K. (2006). From teacher to teacher educator: Experiences, expectations, expatriation. *Studying Teacher Education, 2*(1), 5–23.

Driver, R., Asoko, H., Leach, J., Mortimer, E. & Scott, P. (1994). Constructing scientific knowledge in the classroom. *Educational Researcher, 23*(7), 5–12.

Ducharme, E. R., (1993). *The Lives of Teacher Educators*. New York: Teachers College Press.

Elbaz, F. (1983). *Teacher thinking: A study of practical knowledge*. London: Croom Helm.

Emert, S. C. (1996). *Examining Teacher Thinking Through Reflective Journals: An Educatorís Professional Journey*. Unpublished doctoral thesis, University of Arizona.

Eraut, M. (1994). *Developing professional knowledge and competence*. London: Falmer Press.

Erickson, G. L., & Mackinnon, A. M. (1991). Seeing classrooms in new ways: On becoming a science teacher. In D. A. Schön (Ed.), *The reflective turn* (pp. 15–36). New York: Teachers College Press.

Feiman-Nemser, S., & Remillard, J. (1996). Perspectives on learning to teach. In B. Murray (Ed.), *The Teacher Educator's Handbook: Building a knowledge base for the preparation of teachers* (pp. 63–91). San Francisco: Jossey-Bass.

Fenstermacher, G. D. (1994). The knower and the known: The nature of knowledge in research on teaching. *Review of Research in Education, 20,* 3–56.

Fine, M. (1992). *Disruptive Voices: the possibilities of feminist research.* University of Michigan Press.

Fisher, K. M., & Lipson, J. I. (1986). Twenty questions about student errors. *Journal of Research in Science Teaching, 23*(9), 783–804.

Fitzgerald, L. M., Farstad, J. E., & Deemer, D. (2002). What gets "mythed" in the student evaluations of their teacher education professors? In J. Loughran, & T. Russell, (Eds.), *Improving teacher education practices through self-study* (pp. 208–221). London: Falmer Press.

Freese, A. (2002, April). *Reframing one's teaching: Discovering our teaching selves through reflection and inquiry.* Paper presented at the annual meeting of the American Educational Research Association, New Orleans.

Freire, P. (1986). *Pedagogy of the oppressed.* New York: Continuum.

Fullan, M. (1991). *The new meaning of educational change.* London: Cassell.

Gess-Newsome, J., & Lederman, N. G. (1995). Biology Teachers' Perceptions of Subject Mattter Structure and its Relationship to Classroom Practice. *Journal of Research in Science Teaching 32*(3), 301–325.

Ghaye, T., & Lillyman, S. (1997). *Learning journals and critical incidents: Reflective practice for health care professionals.* Dinton: Quay Books.

Grimmett, P. (1997). Transforming a didactic professor into a learner-focussed teacher educator. In T. R. Carson & D. Sumara (Eds.), *Action research as a living practice* (pp. 121–136). New York: Peter Lang.

Grimmett, P. P., & Chelan, E. P. (1990). Barry: A case of teacher reflection and clinical supervision. *Journal of Curriculum and Supervision, 5,* 214–235.

Guilfoyle, K. (1995). Constructing the Meaning of Teacher Educator: The Struggle to Learn the Roles. *Teacher Education Quarterly, 22*(3), 11–28.

Guilfoyle, K., Hamilton, M. L., Pinnegar, S. & Placier, M. (1995). Becoming teachers of teachers: The paths of four beginners. In T. Russell. & F. Korthagen (Eds.), *Teachers who teach Teachers* (pp. 35–55). London: Falmer Press.

Guilfoyle, K., Hamilton, M. L., & Pinnegar, S. (1997). Obligations to unseen children. In J. Loughran & T. Russell (Eds.), *Teaching about teaching: Purpose, passion and pedagogy in teacher education* (pp. 183–209). London: Falmer Press.

Gunstone, R. F. (2000). Constructivism and learning research in science education. In D. C. Philips (Ed.), *Constructivism in Education: Opinions and Second Opinions on Controversial Issues* (Ninety-ninth Yearbook of the National Society for the Study of Education) (pp.254–280). National Society for the Study of Education / the University of Chicago Press.

Hamilton, M. L. (Ed.) (1998). *Reconceptualising teaching practice: Self-study in Teacher Education.* London: Falmer Press.

Hamilton, M. L. (2004). Professional Knowledge, Teacher Education and Self-Study. In J. J. Loughran, M. L. Hamilton, V. K. LaBoskey, & T. Russell (Eds.), *International Handbook of Self-study of Teaching and Teacher Education Practices* (Vol. 1, pp. 375–419). Dordrecht: Kluwer publishing.

Hamilton, M. L., & Pinnegar, S. (1998a). Introduction: Reconceptualizing teaching practice. In M. L. Hamilton (Ed.), *Reconceptualizing teaching practice: Self-study in teacher education* (pp. 1–4). London: Falmer Press.

Hamilton, M. L., & Pinnegar, S. (1998b). Conclusion: The value and promise of self-study. In M. L. Hamilton (Ed.), *Reconceptualising teaching practice: Self-study in Teacher Education* (pp. 235–246). London: Falmer Press.

Harris, C., & Pinnegar, S. (2000). Using videoethnographies to make the myths of teacher education visible. In J. Loughran & T. Russell (Eds.), *Exploring myths and Legends of Teacher Education.* Proceedings of the Third International Conference of the Self-Study of Teacher Education Practices (pp. 113–117). Herstmonceaux Castle, East Sussex, England. Kingston, Ontario: Queen's University.

Heaton, R., & Lampert, M. (1993). Learning to hear voices: Inventing a new pedagogy of teacher educa-
    tion. In D. K. Cohen, M. W. McLaughlin & J. Talbert (Eds.), *Teaching for understanding* (pp 43–83).
    San Francisco: Jossey-Bass.
Hewson, P. W., Kerby, H. & Cook, P. A. (1995). Determining The Conceptions Of Teaching Science Held
    By Experienced High School Teachers. *Journal of Research in Science Teaching.* 32, (5), 503–520.
Hewson, P. W., Tabachnick, B. R., Zeichner, K. M., Lemberger, J. (1999). Educating prospective teachers
    of biology: Findings, limitations, and recommendations. *Science Education.* 83, (3), 373–384.
Hoban, G. (1997). Learning about learning in the context of a science methods course. In J. Loughran &
    T. Russell (Eds.), *Teaching about teaching: Purpose, passion and pedagogy in teacher education*
    (pp. 133–149). London: Falmer Press.
Hoban, G. F. (2004). Using Information and Communication Technologies for the Self-Study of Teaching.
    In J. J. Loughran, M. L. Hamilton, V. K. LaBoskey, & T. Russell (Eds.), *International Handbook of
    Self-study of Teaching and Teacher Education Practices* (Vol. 2, pp. 1039–1072). Dordrecht: Kluwer
    publishing.
Hoban, G., & Ferry, B. (2001, July). *Seeking the teachable moment in discussions using information and
    communication technologies.* Paper presented at the annual meeting of the Australasian Science
    Education Research Association, Sydney.
Holt-Reynolds, D., & Johnson, S. (2002). Revising the task: The genre of assignment making. In C. Kosnik,
    A. Freese, & A. Samaras (Eds.), *Making a difference in teacher education through self-study.*
    Proceedings of the Fourth International conference on Self-study of Teacher Education Practices,
    Herstmonceux, East Sussex, England (pp. 14–17). Toronto: OISE, University of Toronto.
Holt-Reynolds, D. (2004). Personal history-based beliefs as relevant prior knowledge in course work. In
    In J. J. Loughran, M. L. Hamilton, V. K. LaBoskey, & T. Russell (Eds.), *International Handbook of
    Self-study of Teaching and Teacher Education Practices* (Vol. 1, pp. 343–368). Dordrecht: Kluwer
    publishing.
Hudson-Ross, S., & Graham, P. (2000). Going public: Making teacher educators as a model for preservice
    teachers. *Teacher Education Quarterly, 27*(4), 5–25.
Jersild, A.T. (1955). *When teachers face themselves.* Teachers College Press, Columbia University.
Knowles, J. G., & Holt-Reynolds, D. (1991). Shaping pedagogies through personal histories in preservice
    education. *Teachers College Record, 93* (1), 87–113.
Korthagen, F. A. J. (2004). In search of the essence of a good teacher: towards a more holistic approach in
    teacher education, *Teaching and Teacher Education, 20*(1), 77–97.
Korthagen, F. & Lagerwerf, B. (1996). Reframing the Relationship Between Teacher Thinking and
    Teacher Behaviour: levels in learning about teaching. *Teachers and Teaching: theory and practice,*
    2(2), 161–190.
Korthagen, F. A. J., & Kessels, J. P. A. M. (1996). The relationship between theory and practice: Back to
    the classics. *Educational Researcher*, *25*(3), 17–22.
Korthagen, F. A. J., & Kessels, J. P. A. M. (1999). Linking theory and practice: Changing the Pedagogy of
    Teacher Education. *Educational Researcher, 28*(4), 4–17.
Korthagen, F. J., with Kessels, J., Koster, B., Lagerwerf, B., & Wubbels, T. (2001). *Linking practice and
    theory: The pedagogy of realistic teacher education.* Mahwah, NJ: Erlbaum.
Korthagen, F., Loughran, J., & Lunenberg, M. (2005). Teaching teachers: studies into the expertise of teacher
    educators: an introduction to this theme issue. *Teaching and Teacher Education, 21*(2), 107–115.
Korthagen, F., & Vasalos, A. (2005). Levels in reflection: Core reflection as a means to enhance professional
    growth. *Teachers and Teaching: theory and practice, 11*(1), 47–71.
Kremer-Hayon, L., & Zuzovsky, R. (1995). Themes, processes and trends in the professional development
    of teacher educators. In T. Russell & F. Korthagen (Eds.), *Teachers who teach Teachers* (pp. 155–171).
    London: Falmer Press.
LaBoskey, V. K. (2004). The Methodology of Self-Study and its Theoretical Underpinnings. In J. J. Loughran,
    M.L. Hamilton, V.K. LaBoskey, & T. Russell, (Eds.), *International Handbook of Self-study of Teaching
    and Teacher Education Practices* (Vol. 2, pp. 817–869). Dordrecht: Kluwer publishing.
Lampert, M. (1985). How do teachers manage to teach? Perspectives on problems in practice. *Harvard
    Educational Review, 55*, 178–194.

Lampert, M. (1990). When the problem is not the question and the solution is not the answer. *American Education Research Journal, 27,* 29–63.

Larrivee, B. (2000). Transforming Teaching Practice: Becoming The Critically Reflective Teacher. *Reflective Practice, 1*(3), 293–307.

Lofland, J., & Lofland, L. (1995). *Analysing Social Settings: A Guide to Qualitative Observation and Analysis,* 3rd ed. Belmont, California: Wadsworth.

Lortie, D. C. (1975). *Schoolteacher.* Chicago: Chicago University Press.

Loughran, J. J. (1995). Practising what I preach: Modelling reflective practice to student teachers. *Research in Science Education, 25,* 431–451.

Loughran, J. J. (1996). *Developing reflective practitioners: Learning about teaching and learning through modelling.* London: Falmer Press.

Loughran, J. J. (1997). Teaching about Teaching: Principles and Practice. In J. Loughran & T. Russell (Eds.), *Teaching about teaching: Purpose, passion and pedagogy in teacher education* (pp. 57–70). London: Falmer Press.

Loughan, J. J. (1999). Researching Teaching for Understanding. In John Loughran (Ed.), *Researching Teaching: Methodologies and Practices for Understanding Pedagogy,* (pp. 1–11). London: Falmer Press.

Loughran, J. J. (2002). Effective reflective practice: In search of meaning in learning about teaching. *Journal of Teacher Education, 53*(1), 33–44.

Loughran, J. J. (2004). A History and Context of Self-Study of Teaching and Teacher Education Practices. In J. J. Loughran, M. L. Hamilton, V. K. LaBoskey, & T. Russell (Eds.), *International Handbook of Self-study of Teaching and Teacher Education Practices* (Vol. 1, pp. 7–39). Dordrecht: Kluwer publishing.

Loughran, J. (2005). Researching Teaching about Teaching: Self-study of teacher education practices. *Studying Teacher Education, 1*(1), 5–16.

Loughran, J. J. (2006). *Developing a Pedagogy of Teacher Education: Understanding teaching and learning about teaching.* London: Routledge.

Loughran, J. J. (2007). Science teacher as learner. In Lederman, N., & Abell, S. (Eds.), *Handbook of Science Education.* (pp. 1043–1065). Philadelphia: Erlbaum.

Loughran, J. & Berry, A. (2005). Modelling by teacher educators, *Teaching and Teacher Education,* 21(2), 193–203.

Loughran, J. J., Berry, A., & Corrigan, D. (2001). Once Were Science Teachers. *The Qualitative Report,* 6(4), (December). http://www.nova.edu/ssss/QR/QR6-4/loughran.html

Loughran, J. & Northfield, J. (1996). *Opening the Classroom Door: Teacher, Researcher, Learner.* London: Falmer press.

Loughran, J., & Northfield, J. (1998). A framework for the development of self-study practice. In M. L. Hamilton (Ed.), *Reconceptualizing teaching practice: Self-study in teacher education* (pp. 7–18). London: Falmer Press.

Loughran, J. J., & Russell, T. (1997). Meeting student teachers on their own terms: Experience precedes understanding. In V. Richardson (Ed.), *Constructivist teacher education: Building a world of new understandings* (pp. 164-181). London: Falmer Press.

Louie, B. Y., Stackman, R. W., Drevdahl, D. & Purdy, J. M. (2002). Myths about teaching and the university professor. The power of unexamined beliefs. In J. Loughran & T. Russell (Eds.), *Improving teacher education practices through self-study* (pp. 193–207). London: RoutledgeFalmer Press.

Lunenberg, M. (2002). Designing a Curriculum for Teacher Educators. *European Journal of Teacher Education,* 25 (2), 263–27.

Lyons, N. (1990). Dilemmas of knowing: Ethical and epistemological dimensions of teachers' work and development. *Harvard Educational Review, 60,* 159–180.

McDiarmid, G. W. (1990). Challenging prospective teachers during early field experiences: A quixotic undertaking? *Journal of Teacher Education, 41*(3), 12–20.

McNiff, J. (1996). The Other Side of Silence: Towards a Theory of Listening. In J. Richards & T. Russell (Eds.), *Empowering our future in teacher education.* Proceedings of the First International Conference on Self-Study of Teacher Education Practices, Herstmonceux Castle, East Sussex, England (pp. 1–4). Kingston, Ontario: Queen's University.

McNiff, J., Lomax, P., & Whitehead, J. (1996). *You and your action research project.* London: Routledge.

Macgillivray, L. (1997). Do what I say, not what I do: An instructor rethinks her own teaching and research. *Curriculum Inquiry, 24*(4), 469–488.

Mason, J. (2002). *Researching your own practice. The Discipline of Noticing.* London: Routledge Falmer.

Mayeroff, M. (1971). *On caring.* New York: Harper & Row.

Mayes, C. (2001). A transpersonal model for teacher reflectivity. *Journal of Curriculum Studies, 33*(4), 447–493.

Measor, L. (1985). Critical Incidents in the Classroom: Identities, Choices and Careers. In Ball, S. J. and Goodson, I. F. (Eds.), *Teachers' Lives and Careers* (pp. 61–77). London: Falmer Press.

Mitchell, I. (1999). Bridging the Gulf Between Research and Practice. In Loughran, J. (Ed.), *Researching Teaching: Methodologies and Practices for Understanding Pedagogy* (pp. 44–64). London: Falmer press.

Mitchell, C., & Weber, S. (1999). *Reinventing ourselves as teachers: Beyond Nostalgia.* London: Falmer Press.

Mueller, A. (2001, April). *Reflecting with teacher candidates on the complexities and wonders of learning to teach.* Paper presented at the meeting of the American Educational Research Association Conference, Seattle.

Munby, H. & Russell, T. (1992). Transforming chemistry research into chemistry teaching: The complexities of adopting new frames for experience. In T. Russell & H. Munby (Eds.), *Teachers and teaching: From classroom to reflection* (pp. 90–123). London: Falmer Press.

Munby, H., & Russell, T. (1994). The authority of experience in learning to teach: Messages from a physics method class. *Journal of Teacher Education, 45*(2), 86–95.

Munby, H., & Russell, T. (1995). Towards Rigour with Relevance: How can Teachers and teacher Educators Claim to Know? In T. Russell & F. Korthagen (Eds.), *Teachers who teach Teachers* (pp. 172–186). London: Falmer Press.

Munby, H., Russell, T., & Martin, A. K. (2001). Teachers' knowledge and how it develops. In V. Richardson (Ed.), *Handbook of research on teaching* (pp. 877–904). Washington, DC: American Educational Research Association.

Murray, J. (2005). Re-addressing the priorities: new teacher educators and induction into higher education. *European Journal of Teacher Education, 28*(1), 67–85.

Myers, C. B. (2002). Can self-study challenge the belief that telling, showing and guided-practice constitute adequate teacher education? In J. Loughran & T. Russell (Eds.), *Improving teacher education practices through self-study* (pp.130–142). London: RoutledgeFalmer.

Nicol, C. (1997a). *Learning to teach prospective teachers to teach mathematics.* Unpublished doctoral dissertation, University of British Columbia, Vancouver.

Nicol, C. (1997b). Learning to Teach Prospective Teachers to Teach Mathematics: The Struggles of a Beginning teacher Educator. In John Loughran & Tom Russell (Eds.), *Teaching about Teaching: Purpose, Passion and Pedagogy in Teacher Education* (pp. 95–116). London: Falmer Press.

Noddings, N. (2001). The caring teacher. In V. Richardson (Ed.), *Handbook of research on teaching* (pp. 99–105). Washington, DC: American Educational Research Association.

Northfield, J., & Loughran, J. (1996). Learning through self-study: Exploring the development of knowledge. In J. Richards & T. Russell (Eds.), *Empowering our future in teacher education.* Proceedings of the First International Conference on Self-Study of Teacher Education Practices, Herstmonceux Castle, East Sussex, England (pp. 180–182). Kingston, Ontario: Queen's University.

Northfield, J. & Gunstone, R. (1997). Teacher Education as a Process of Developing Teacher Knowledge. In J. Loughran & T. Russell (Eds.), *Teaching about Teaching: Purpose, Passion and Pedagogy in Teacher Education* (pp. 48–56). London: Falmer Press,

Pajares, M. (1992). Teachers' beliefs and educational research: Cleaning up a messy construct. *Review of Educational Research, 62,* 307–332.

Palmer, P. J. (1998). *The Courage to Teach: Exploring the Inner Landscape of a Teacher's Life.* Jossey-Bass: San Francisco.

Patton, M. Q. (2002). *Qualitative Research and Evaluation Methods.* (3rd ed.) Thousand Oaks/London, Sage.

Peterman, F. (1997). The lived curriculum of constructivist teacher education. In V. Richardson, (Ed.), *Constructivist teacher education: Building a world of new understanding* (pp. 154–163). London: Falmer Press.

Pinnegar, S. (1998). Introduction to Part II: Methodological perspectives. In M. L. Hamilton (Ed.), *Reconceptualising teaching practice: Self-study in Teacher Education* (pp. 30–33). London: Falmer Press.

Polyani, M. (1966). *The tacit dimension.* Garden City, NY: Doubleday.

Pope, C. A. (1999). Reflection and refraction: A reflexive look at an evolving model for methods instruction. *English Education, 31*(3), 177–201.

Posner, G. J., Strike, K. A., Hewson, P. W., & Gertzog, W. A., (1982). Accomodation of a scientific conception: Towards a theory of conceptual change. *Science Education, 66*(2), 211–227.

Richards, J. C. (1998). Turning to the Artistic: Developing an Enlightened Eye by Creating Teaching Self-Portraits. In Mary Lynn Hamilton, (Ed.), *Reconceptualizing teaching practice: Self-study in teacher education* (pp. 34–44). London: Falmer Press.

Ritter, J. K. (2006). The difficulties of forging a teacher educator pedagogy: transitioning from classroom teacher to teacher educator. In L. M. Fitzgerald, M. L. Heston, and D. L. Tidwell (Eds.), *Collaboration and Community: Pushing Boundaries through Self-Study.* Proceedings of the Sixth International Conference on Self-Study of Teacher Education Practices (pp. 216–219). Cedar Falls, IA: University of Northern Iowa.

Rogers, C. (1969). *Freedom to Learn.* Columbus, OH: Merrill.

Rowe, M. B. (1974a). Wait-Time and rewards as instructional variables, their influence on language, logic, and fate control: part one - wait-time. *Journal of Research in Science Teaching, 11*(2), 81–94.

Rowe, M. B. (1974b). Relation of wait-time and rewards to the development language, logic and fate control: part two: rewards. *Journal of Research in Science Teaching, 11*(4), 291–308.

Russell, T. (1997). Teaching teachers: How I teach IS the message. In J. Loughran & T. Russell (Eds.), *Teaching about teaching: Purpose, passion and pedagogy in teacher education* (pp. 32–47). London: Falmer Press.

Russell, T. (1999). The Challenge of change in teaching and teacher education. In J.R. Baird (Ed.), *Reflecting, teaching, learning: Perspectives on educational improvement* (pp. 219–238). Cheltenham, Victoria: Hawker Brownlow Education.

Russell, T., & Bullock, S. (1999). Discovering Our Professional Knowledge as teachers: Critical Dialogues about Learning from Experience. In J. Loughran (Ed.), *Researching Teaching: Methodologies and Practices for Understanding Pedagogy* (pp. 132–151). London: Falmer Press.

Russell, T., & Loughran, J. (2007). *Enacting a Pedagogy of Teacher Education: Values, Relationships and Practices.* London: Routledge.

Russo, P. & Beyerbach, B. (2001). Moving from Polite Talk to Candid Conversation: Infusing Foundations into a Professional Development Project. *Educational Foundations, 15*(2), 71–90.

Sanjek, R. (1990). A Vocabulary for Field Notes. In R. Sanjek (Ed.), *Field Notes: The Making of Anthropology.* Ithaca, New York: Cornell University Press.

Schön, D. A. (1983). *The reflective practitioner: How professionals think in action.* New York: Basic Books.

Schön, D. A. (1987). *Educating the reflective practitioner.* San Francisco: Jossey-Bass.

Schulte, A. (2001). *Student teachers in transformation: A self-study of a supervisor's practice.* Unpublished doctoral dissertation, University of Wisconsin, Madison.

Segal, A. (2002). *Disturbing Practice: Reading Teacher Education As Text.* Peter Lang: New York.

Senese, J. C. (2002). Opposites attract. What I learnt about being a classroom teacher by being a teacher educator. In J. Loughran & T. Russell (Eds.), *Improving teacher education practices through self-study* (pp. 43–55). London: Falmer Press.

Stets, J. E., & Burke, P. J. (2000). Identity Theory and Social Identity Theory. *Social Psychology Quarterly, 63*(3), 224–237.

Stronach, I., Corbin, B., McNamara, O., Stark, S., & Warne, T. (2002). Towards an uncertain politics of professionalism: teacher and nurse identities in flux. *Journal of Education Policy, 17*(1), 109–138.

Tidwell, D. (2002). A balancing act: Self-study in valuing the individual student. In J. Loughran & T. Russell (Eds.), *Improving teacher education practices through self-study* (pp. 30–42). London: Falmer Press.

Tidwell, D., & Fitzgerald, L. (Eds.) (2006). *Self-study and Diversity*. Sense. Rotterdam. The Netherlands.

Trumbull, D. J. (1999). *The New Science Teacher*. Teachers College Press. New York and London.

Trumbull, D. J. (2000). Comments to students and their effects. In J. Loughran & T. Russell (Eds.), *Exploring myths and legends of teacher education*. Proceedings of the Third International Conference on the Self-Study of Teacher Education Practices, Herstmonceux Castle, East Sussex, England (pp. 243–247). Kingston, Ontario: Queen's University.

Trumbull, D. (2004). Factors important for the scholarship of self-study of teaching and teacher education practices. In J. J. Loughran, M. L. Hamilton, V. K. LaBoskey, & T. Russell, (Eds.), *International Handbook of Self-study of Teaching and Teacher Education Practices* (Vol. 2, pp. 1211–1230). Dordrecht: Kluwer publishing.

van Manen, M. (1990). *Researching lived experience*. London, Ontario: Althouse Press.

van Manen, M. (1991). Reflectivity and the pedagogical moment: the normativity of pedagogical thinking and acting. *Journal of Curriculum Studies, 23*(6), 507–536.

Warren Little, J., Gearhart, M., Curry, M., & Kafka, J. (2003). Looking at Student Work For Teacher Learning, Teacher Community, and School Reform. *Phi Delta Kappan, 85*(3), 185–192.

Wideen, M., Mayer-Smith, J., & Moon, B. (1998). A critical analysis of the research on learning to teach: Making the case for an ecological perspective on inquiry. *Review of Educational Research, 68*(2), 130–178.

Wiersma, W. (1986). *Research Methods in Education: An Introduction*. (4th ed.) Boston: Allyn and Bacon.

White, B. C. (2002). Constructing constructivist teaching: Reflection as research. *Reflective Practice, 3*, 307–326.

White, R. T. (1988). *Learning science*. London: Blackwell.

White, R. T., & Gunstone, R. F. (1992) *Probing Understanding*. Falmer.

Whitehead, J. (1993). *The growth of educational knowledge: Collected papers*. Bournemouth: Hyde Publications.

Whitehead, J. (1998). How do I know that I have influenced you for good? A question of representing my educative relationships with research students. In A. L. Cole & S. Finley (Eds.), *Conversations in community*. Proceedings of the Second International Conference of the Self-Study of Teacher Education Practices, Herstmonceux Castle, East Sussex, England (pp. 10–12). Kingston, Ontario: Queenís University.

Wideen, M., Mayer-Smith, J., & Moon, B. (1998). A critical analysis of the research on learning to teach: Making the case for an ecological perspective on inquiry. *Review of Educational Research, 68*(2), 130–178.

Wilkes, G. (1998). Seams of paradoxes in teaching. In M. L. Hamilton, (Ed.), *Reconceptualising teaching practice: Self-study in teacher education* (pp. 198–207). London: Falmer Press.

Wilson, S. (2006). Finding a canon and core: Meditations on the preparation of teacher educator-researchers. *Journal of Teacher Education, 57*(3), 315–325.

Winter, R. (2006). What Do You Do in Library Method: Learn to Say "Ssshhhh!!"?. In L. M. Fitzgerald, M. L. Heston, and D. L. Tidwell (Eds.), *Collaboration and Community: Pushing Boundaries through Self-Study*. Proceedings of the Sixth International Conference on Self-Study of Teacher Education Practices (pp. 272–275). Cedar Falls, IA: University of Northern Iowa.

Zeichner, K. (1999). The new scholarship in teacher education. *Educational Researcher, 28*(9), 4–15.

Zeichner, K. M. (2005). Becoming a teacher educator: a personal perspective. *Teaching and Teacher Education, 21*(2), 117–124.

Zeichner, K., & Gore, J.M. (1990). Teacher socialization. In W. R. Houston (Ed.), *Handbook of research on teacher education* (pp. 329–348). New York: Macmillan.

SUBJECT INDEX

AUTHOR INDEX

Self Study of Teaching and Teacher Education Practices (Series)

1. G. Hoban (ed.): *The Missing Links in Teacher Education Design.* Developing a Multi-linked Conceptual Framework. 2005   ISBN: 1-4020-3338-9
2. C. Kosnik, C. Beck, A.R. Freese and A.P. Samaras (eds.): *Making a Difference in Teacher Education Through Self-Study.* Studies of Personal, Professional and Program Renewal. 2006   ISBN: 1-4020-3527-6
3. P. Aubusson and S. Schuck (eds.): *Teacher Learning and Development.* The Mirror Maze. 2006   ISBN: 1-4020-4622-7
4. L.F. Darling, G. Erickson and A. Clarke (eds.): *Collective Improvisation in a Teacher Education Community.* 2007   ISBN: 1-4020-5667-2
5. Amanda Berry: *Tensions in Teaching about Teaching.* 2007
          ISBN: 978-1-4020-5992-6

springer.com

Printed in the United States
77309LV00003BB/13-51

DATE DUE

Demco, Inc. 38-293